The DeepSeek Moment

A New Era of Efficiency, Intelligent Design, and Democratization

Harshad Oak

Monish Darda

Apress®

The DeepSeek Moment: A New Era of Efficiency, Intelligent Design, and Democratization

Harshad Oak
Pune, Maharashtra, India

Monish Darda
Pune, Maharashtra, India

ISBN-13 (pbk): 979-8-8688-2597-2
https://doi.org/10.1007/979-8-8688-2598-9

ISBN-13 (electronic): 979-8-8688-2598-9

Managing Director, Apress Media LLC: Welmoed Spahr
Acquisitions Editor: Shivangi Ramachandran
Development Editor: James Markham
Project Manager: Jessica Vakili

Distributed to the book trade worldwide by Springer Science+Business Media New York, 1 New York Plaza, New York, NY 10004. Phone 1-800-SPRINGER, fax (201) 348-4505, e-mail orders-ny@springer-sbm.com, or visit www.springeronline.com. Apress Media, LLC is a Delaware LLC and the sole member (owner) is Springer Science + Business Media Finance Inc (SSBM Finance Inc). SSBM Finance Inc is a **Delaware** corporation.

For information on translations, please e-mail booktranslations@springernature.com; for reprint, paperback, or audio rights, please e-mail bookpermissions@springernature.com.

Apress titles may be purchased in bulk for academic, corporate, or promotional use. eBook versions and licenses are also available for most titles. For more information, reference our Print and eBook Bulk Sales web page at http://www.apress.com/bulk-sales.

If disposing of this product, please recycle the paper

Table of Contents

Chapter 9: The Agentic Shift and the Emergence of Service-as-a-Software

About the Authors

Harshad Oak is a technologist, entrepreneur and author with over two decades of experience building enterprise software products. He has founded IndicThreads and Rightrix Solutions and held senior leadership roles at Oracle and Icertis. He is the author of four previous technology books and has spoken at conferences worldwide on enterprise software and applied AI. Recognized as a Java Champion and an Oracle ACE Director, he holds a Master's degree in Software Development from Symbiosis and an MBA from the Indian School of Business (ISB). Beyond his professional work, he actively supports science education and rationalist causes in India.

linkedin.com/in/harshadoak/

Monish Darda is the co-founder of Humantronik and Icertis, where he continues to serve as Chief Mentor. With over three decades of experience building large-scale enterprise software, he has been at the forefront of enterprise technology innovation. In 2026, he founded Humantronik to enable enterprises to convert human intent into measurable business outcomes through personalized AI. At Icertis, he helped forge the world's leading Contract Intelligence platform – technology powerful enough to steward millions of agreements and over a trillion dollars in value. He is the co-inventor of four AI patents and two patents in cloud resource management and provisioning. Monish holds a BE in Mechanical Engineering and a master's degree from Florida Atlantic University. His passions include reading, gaming, bikes, and cars, and he is a confirmed gadget freak.

linkedin.com/in/monishdarda/

Foreword

The DeepSeek Moment

Every generation of technology has moments when the direction of innovation quietly shifts. At first, these moments are visible only to a few practitioners paying close attention. Over time, however, they reshape industries and redistribute opportunity. We may now be witnessing such a moment in artificial intelligence, a shift from AI being dominated by a handful of companies toward ingenuity, efficiency, and broader participation.

I have known Monish Darda for more than 25 years. In Pune's technology community, he has always stood out as a rare combination—a deep technologist with a sharp business sense. That combination is far less common than people imagine. Monish has consistently demonstrated that while understanding technology deeply and understanding how businesses work are two different skills, he has excelled in both.

Over the years, I have also noticed something about the way he thinks. Most technologists ask, "How do we build this?" Monish often asks a different question: "Why are we building it this way?" Those questions tend to unsettle comfortable assumptions, but they are also the ones that move industries forward. That instinct runs through every chapter of this book.

I first met Harshad Oak when he was running IndicThreads. What struck me immediately was that he understood something fundamental about technology ecosystems: ideas do not spread on technical merit alone. They spread through communities, networks, and conversations. Harshad has always been someone who connects people and ideas across the industry, and that perspective shows clearly in this book. It draws from many voices and experiences across the technology landscape.

When Monish and Harshad asked me to write this foreword, I did not hesitate. I respect them for their technical depth, their curiosity, and the fact that they have remained engaged with technology through decades of rapid change. That curiosity is important. In fields like ours, the moment you believe you have fully understood the system, the system changes.

Over the last 35 years, I have watched several waves of computing unfold. Each of these shifts has followed a similar pattern. The early phase is dominated by a few large players with the capital to build the infrastructure. Over time, innovation moved outward as tools became cheaper and more accessible. What we may be witnessing with DeepSeek is another such transition—a move from scale alone to efficiency and ingenuity.

The event that triggered this book was dramatic enough to capture global attention. On January 27, 2025, news related to DeepSeek triggered a historic reaction in financial markets and wiped hundreds of billions of dollars from Nvidia's market value in a single day. Markets react quickly, sometimes emotionally. But what mattered was not the market movement itself. What mattered was what it revealed.

What DeepSeek demonstrated was simple but important. Frontier AI capability does not always belong to the organization with the largest GPU cluster. Sometimes it belongs to the team that designs the system more intelligently.

For several years, the dominant narrative in artificial intelligence was simple: bigger models, more data, more compute. Progress appeared to belong almost exclusively to a handful of extremely well-funded companies. In such a world, most organizations were effectively renting their AI future rather than building it.

This book challenges that assumption.

Darda and Oak walk the reader through how we arrived at this moment and why the DeepSeek breakthrough matters. They explain the technical context clearly, but they also go beyond technology. The book looks at what this shift means for enterprises trying to adopt AI, for policymakers trying to understand its implications, and for countries thinking about their place in the emerging AI landscape.

One line from the book stayed with me: "awe gets you applause, but ROI gets you budgets." Anyone who has spent time inside large enterprises knows how true that is. AI may inspire fascination, but sustainable adoption happens only when organizations learn to deploy it with discipline and measurable outcomes.

The book also raises an idea that I find particularly powerful: the analogy with Gutenberg. The printing press did not simply make books cheaper. It redistributed knowledge and shifted power away from a small elite. Something similar may now be happening with AI. If intelligence becomes dramatically cheaper and more accessible, the center of innovation will shift.

The constraint will no longer be access to compute alone. The constraint will increasingly be imagination, creativity, and execution.

The final chapters explore what this means for the future of software. The authors discuss the emergence of agentic systems and what they call Service-as-a-Software. The idea is provocative but plausible. When reasoning becomes inexpensive and abundant, software will begin to behave less like static applications and more like adaptive systems capable of acting on behalf of users.

Technology moments matter most when they inspire people to create.

Moments like this create openings for new participants in the technology ecosystem. India is well-positioned to take advantage of such moments. Over the past three decades our engineers have built and operated some of the most complex software systems in the world. That experience matters. The next step is to apply that capability to building AI systems that are rooted in India's realities—our languages, our data, our scale, and our societal challenges.

Efforts such as Sarvam and BharatGen are encouraging precisely because they represent this ambition. They are early efforts, and the work ahead is difficult. But they point toward the direction India must pursue: building, not merely consuming.

The redistribution of intelligence that this book describes is not guaranteed to benefit everyone equally. It will reward those who invest in research, build strong ecosystems, and have the courage to experiment.

The authors close the book with a simple question: What will you build?

It is the right question. And this book earns the right to ask it.

In a field filled with excitement, hype, and sometimes confusion, Darda and Oak have written something that is thoughtful, grounded, and genuinely useful. They explain the past clearly, analyze the present honestly, and encourage readers to think carefully about what comes next.

If you are a technologist, entrepreneur, enterprise leader, or policymaker trying to understand the forces shaping AI today, you will find this book worth your time.

And if you read it carefully, you may also find yourself asking that same question.

What will you build?

Anand Deshpande

Founder & Chairman

Persistent Systems Limited

Pune, India

Preface

Science fiction stories often rely on the premise of abundant, unlimited energy to build their utopian futures. We are not there with energy yet. However, we are rapidly arriving at a reality of abundant intelligence. The marginal cost of reasoning is collapsing.

As foundational AI models evolved, the technology industry assumed that building advanced artificial intelligence required massive data centers and billions of dollars. That assumption created an elite class of companies that had the rights to the foundation models. Then came the DeepSeek breakthrough. A relatively small team proved that architectural elegance could outmaneuver brute-force capital. They shattered the myth that high-level cognitive computing was a luxury good. It is now a basic, accessible building block available to anyone with the curiosity and discipline to use it.

We might not fully see the magnitude of this shift while we are busy living through it. But future generations will look back on this exact moment. They will view the democratization of reasoning the exact same way we view the democratization of knowledge with the printing press or the democratization of industry and manufacturing with the steam engine. It is the sudden unlocking of a fundamental constraint on human progress.

Both of us have spent our careers building complex software for some of the largest and most demanding organizations in the world. Through that experience, we have understood the stark difference between a fascinating pilot project and a resilient, trusted enterprise solution. We have lived through the pains of adopting a new technology. Now that the cost of reasoning is falling rapidly, the traditional limitations on what we can build, how quickly, and how cheaply are permanently removed.

Some leaders and organizations are thriving in these times of change, but many more are overwhelmed by it. It is similar to asking someone what they would do if they suddenly received a billion dollars. Most people freeze after a few initial ideas. They know they could do remarkable things with the money, but they somehow end up only thinking of buying cars and houses. They do not know where to begin.

Such are the times we live in today. Each of us now has more raw intelligence available at our fingertips than the most powerful people in the world or even the governments of countries had just a few years back.

We wrote this book to provide a clear, optimistic, and highly practical guide to this transition. We want to move the conversation away from anxiety and toward agency.

In the chapters ahead, we move past the generic hype to explore the mechanics of this shift. We examine how smart organizations are using technology to build real, lasting value. We will discuss the mindset needed to navigate the critical challenges of ethics and accountability. In a world where raw intelligence is cheap, systems that can be trusted will win. We will share frameworks to ensure the products you build or use remain safe and reliable.

We will also map the shifting global realities. The DeepSeek moment proved that innovation is no longer concentrated solely in Silicon Valley. As the geopolitical map of technology is redrawn, new superpowers are emerging outside the traditional strongholds.

Finally, we will explore the agentic shift. We will discuss the mental models required to lead in a world where software executes tasks independently. We hope you will find some great, practical tips to design systems that seamlessly blend human oversight and machine autonomy.

Now is the time to take bold action. Artificial intelligence is no longer scarce. The winners in the coming decade will be those who treat it as foundational infrastructure. We hope this book will equip you for this new reality and empower you to build what comes next.

Capturing a technological revolution on paper is an intense process, and we did not do it alone. We are deeply grateful to the engineers, product managers, and enterprise leaders who have shared their journeys with us. Their triumphs, frustrations, and innovations provided the raw material for these chapters. We also thank our families for their unwavering support and immense patience.

We would like to thank Manoj Padki, Kabir Narang, Sanket Atal and Mark Bergen for taking the time to share their insightful comments, which truly helped make this a better book. We also want to thank Anand Deshpande for writing the book's foreword. We are certain you will find his sharp insights beneficial and a great segue into the book content.

The team at Apress and Springer brought expertise, rigor, and patience to a manuscript about a world that refused to stand still while we were writing about it. We are grateful for their partnership and their faith in this project from the very beginning. Special thanks to our Acquisitions Editor Shivangi for believing in this book and Project Manager Jessica for keeping it on track. And to James, Susan, Krishnan, Dulcy, and Joseph, whose craft turned the manuscript into a book.

We invite you to turn the page. We will begin by exploring how the old rules were broken, so that we can understand how to write the new ones.

Harshad Oak, Monish Darda

PS: As this book went to press in late April 2026, DeepSeek released two open-weight models (V4-Pro and V4-Flash) with a 1-million-token context, MIT licensing, and trained partly on Chinese Huawei Ascend chips rather than Nvidia alone. V4 confirms the story we tell in this book. The DeepSeek Moment was never about a single model release; it was about the fall of the doctrine that the future of intelligence belonged to a few. V4 is another swing of the hammer at a wall that DeepSeek R1 first cracked.

CHAPTER 1

The Reign of the Giants

For the CEO of Quantyv Nexis,[1] a leader in data analytics, the question on the top of her mind before the January 2025 board meeting wasn't "What is our AI strategy?" but, rather, "How long until we're just a consumer of someone else's?"

For months, she'd felt trapped. Her company, like thousands of others, was caught in the gravitational pull of the artificial intelligence (AI) giants. They were paying astronomical fees to OpenAI, Google, and Microsoft for a ticket to stay relevant, but the investment wasn't paying off. The investment was delivering some results, but with each small win, the dependency grew, and the costs mounted. The path of building their own capabilities felt impossibly steep.

The prevailing wisdom was clear: you either paid the giants for access to their massive models or you risked being subsumed by them. Her company wasn't building an AI future; it was renting one.

The myth that only a handful of trillion-dollar companies could forge the future of intelligence seemed unbreakable. Then, on January 27, 2025, the myth shattered.

---

[1] Quantyv Nexis is a fictional name used to represent common industry experiences.

© Harshad Oak, Monish Darda 2026

H. Oak and M. Darda, *The DeepSeek Moment*, https://doi.org/10.1007/979-8-8688-2598-9_1

A news alert flashed across her screen:

(Reuters) "Global investors dumped tech stocks on Monday as they worried that the emergence of a low-cost Chinese artificial intelligence model would threaten the dominance of AI leaders like Nvidia, evaporating $593 billion of the chipmaker's market value, a record one-day loss for any company on Wall Street."

Since 2022, the technology industry had been consumed by an artificial intelligence gold rush. The frantic race to buy digital pickaxes and shovels from a handful of suppliers was finally being challenged. The doctrine of "bigger AI is better AI" was facing its first real test, and for millions of business leaders, it was a long-overdue moment of hope.

But before we explore that change, let's understand how the "bigger is better" belief came to rule the world.

The Spark: A New Kind of Technology

The AI gold rush began not with a single event, but with a fundamental technological shift. For years, AI and Machine Learning (ML) simmered quietly. Most companies then saw AI as something you built yourself for a specific use case, using your own clean, proprietary data. It was difficult, expensive, and narrow.

The old way of building AI was time-consuming. Models learned from carefully labeled data, with the labeling process being a major bottleneck that limited the scale and scope of what AI could do.

And the bottleneck was people—you had to have an army to label data in quantities that were statistically significant. At one time, Icertis employed over 300 paralegals to help curate and label data for its AI models and created millions of hand-crafted data points that helped our models perform better.

The breakthrough came in 2017 from a team of eight researchers seven of whom worked at Google and one was with the University of Toronto. In a landmark paper titled "Attention Is All You Need," they introduced the Transformer architecture. Vaswani, Shazeer, Parmar, Uszkoreit, Jones, Gomez, Kaiser, and Polosukhin introduced a revolutionary technique that enabled the model to learn from the raw, unlabeled text of the entire internet. It learned grammar, context, and nuance by predicting the next word in a sentence or filling in blanks.

Coupled with the raw power of Graphics Processing Units (GPUs), which excel at training computations because of their architecture and compute advantages over traditional CPUs, this unlocked the ability to train on unfathomably large datasets, laying the foundation for the massive models to come.

The Stampede: From Lab to Global Phenomenon

Although the Transformer was the spark, OpenAI's ChatGPT in November 2022 truly lit the AI fire. OpenAI had released GPT before the launch of ChatGPT, but that required calling an API and some programming skills. ChatGPT wrapped the power of GPT into a simple chat window that anyone could use and appreciate.

This transformed the landscape. It made the potential of AI obvious not just to technologists but also to the masses, the media, and, most importantly, investors with deep pockets. The idea of a general-purpose AI was no longer theoretical. With ChatGPT, everybody sensed this was a technology that had the potential of creating an opportunity bigger than humanity had ever seen before. As the industry began to pursue Artificial General Intelligence, which meant creating machines capable of passing the Turing test and matching or exceeding human-level intellect, the gold rush had officially started.

The Arms Race: The Gospel of "Bigger Is Better"

The masters of this new world, the ones selling the maps to it, were the giants: Google, OpenAI, Amazon, and Microsoft (and others backed by governments or with access to billions). The principle was simple and seductive: the bigger your mining operation, the more gold you would find.

Success became a function of brute force, fueling an arms race of staggering proportions. The models were now being trained on billions of parameters. You can think of parameters as a tunable knob or an internal variable that the model adjusts during training. The more knobs, the more nuance the model can learn.

To understand parameters, consider whether a cab driver would be better at choosing the best path if he has driven from point A to B three times or if he has done it 300 times with variations in traffic, road speeds, and time of day.

After 300 rides, it would be hard even for the driver to tell why he thinks a certain route is ideal for a certain combination of traffic, day, weather, and other conditions, but he would be able to navigate much better. It works because for every ride, some parameters within that person are getting tweaked and are learning from each experience—something akin to what humans call experience or even intuition.

Similarly, large language models (LLMs) that are trained on lots of good (and bad) data keep adjusting their internal parameters until they get exceptionally good at predicting the next word, or what's known as a token. We will look into this in greater detail in a later chapter.

The whole AI arms race was based on the idea that you wanted your cab driver (AI) to have processed more good data than bad data on thousands of varied journeys on a route, so that it gets brilliant at making predictions.

The belief was that your AI was only as good as the amount of data it had been trained on, which meant needing access to expensive data, expensive compute, and even expensive talent that knew how to handle and manage such mega training exercises.

Size Matters

You've likely seen visualizations of the universe on National Geographic or YouTube. They often begin with a human and then zoom out to the size of Planet Earth, the Sun, and the Milky Way. Every time the visualization expands, you see that what seemed huge a moment ago is tiny in the next frame. The Earth is dwarfed by the Sun, which is just one of billions of stars in our galaxy.

Something similar was happening with LLM parameter sizes. With every new release, the industry would gasp at a model's scale, only for the next one to dwarf it completely. Success became all about brute force. An arms race kicked off, and the only score that seemed to matter was your model's "parameter count."

The escalation was breathtaking. GPT-2 set the stage with 1.5 billion parameters. A few years later, models like OpenAI's GPT-4, Google Gemini, and others were estimated to have parameter counts in the trillions, reportedly costing upward of $100 million to train.[2] We were discussing the capital required to build a mega-factory from the ground up (see Table 1-1).

---

[2] https://www.visualcapitalist.com/training-costs-of-ai-models-over-time

Table 1-1. *The Disruption Landscape (January 2025)*

Model	Release	Parameters	Strategic Focus/Market Role[3]
1 GPT-4o	May 2024	Undisclosed	The "safe" default. Dominant in speed and multimodal (voice/image) features, but criticized for high costs.
2 DeepSeek-R1[4]	January 2025	671B (37B active)	Matches OpenAI o1 reasoning performance at a fraction of the inference cost.
3 Gemini 1.5 Pro[5]	February 2024	Undisclosed	Differentiated by its large context window of 1M+ tokens, Gemini now had the ability to process entire books or codebases.
4 Llama 3.1 405B[6]	July 2024	405 billion	Proof that open-weight models could match the leading proprietary models.
5 Claude 3.5 Sonnet[7]	June 2024	Undisclosed	Widely considered the best model for coding capability.
6 Mistral Large 2[8]	July 2024	123 billion	The European Contender. A balance of performance and size.

---

[3] https://openai.com/index/hello-gpt-4o/

[4] https://github.com/deepseek-ai/DeepSeek-R1

[5] https://storage.googleapis.com/deepmind-media/gemini/gemini_v1_5_report.pdf

[6] https://ai.meta.com/research/publications/the-llama-3-herd-of-models/

[7] https://www-cdn.anthropic.com/fed9cc193a14b84131812372d8d5857f8f304c52/Model_Card_Claude_3_Addendum.pdf

[8] https://mistral.ai/news/mistral-large-2407/

To understand the magnitude of the "DeepSeek moment," we must look at the battlefield as it stood on the morning of January 20, 2025, the day DeepSeek-R1 was released. The industry was dominated by a belief that intelligence required massive scale and massive secrecy. The "giants" were locked in an arms race to build the largest, most expensive black boxes on Earth. Table 1-1 captures this moment of disruption.

While the incumbent giants competed on scale and capability, DeepSeek-R1 entered the market with a radical focus on efficiency, achieving top-tier reasoning with significantly lower active parameters.

Note the contrast in the "Parameters" column. By 2025, the incumbent giants had largely stopped disclosing the size of their models, treating efficiency as a trade secret. DeepSeek broke this silence. By publishing their architecture, they revealed that they were not just matching the giants; they were doing so with a fraction of the active compute.

The Walls of the Walled Garden

This "size doctrine" created huge barriers to entry, built on a few critical resources. The first wall was the sheer cost and effort involved in the infrastructure and hardware setup. Before a company could even think about GPUs, it needed a specialized data center to house them. Building (or renting where possible) these facilities, with their immense power and cooling requirements, was a multimillion-dollar investment in itself.

The energy challenge was so fundamental that OpenAI CEO Sam Altman had suggested that future AI progress requires a breakthrough, like nuclear fusion, just to power the models of tomorrow.[9]

---

[9] https://www.reuters.com/technology/openai-ceo-altman-says-davos-future-ai-depends-energy-breakthrough-2024-01-16

This new level of ambition was evident in proposals such as "Stargate," a joint venture involving OpenAI, SoftBank, Oracle, and MGX. The project's goal was to invest up to $500 billion by 2029, aiming to create the largest computing infrastructure projects in history.[10]

But even with a fortress built, you still had to fight for the components to put inside. The GPU supply chain was so constrained at the time that Oracle's cofounder, Larry Ellison, recounted how he and Elon Musk took Nvidia's CEO Jensen Huang to dinner and literally had[11] to "beg" for more chips.

The second wall was talent. The handful of researchers and engineers who could build and train these colossal models were in extraordinarily high demand. They were quickly hired by the tech giants, who could offer compensation packages and access to computational resources that startups and academic institutions simply couldn't match.

A Future Ruled by Giants?

The result was a growing fear across the industry, not just about the cost, but about control. The narrative was set: the future of AI seemed to belong exclusively to those with the deepest pockets and the biggest data centers.

This consolidation raised concerns that a handful of powerful companies would get to decide the future for everyone else. It felt as though the keys to the next technological revolution were being locked away. But this capital-intensive strategy, while creating monsters of capability, was also planting the seeds of its own undoing.

---

[10] https://techcrunch.com/2025/01/21/openai-teams-up-with-softbank-and-oracle-on-50b-data-center-project

[11] https://fortune.com/2024/09/16/larry-ellison-elon-musk-begged-nvidias-jensen-huang-more-gpus-fancy-sushi-dinner

CHAPTER 2

The DeepSeek Breakthrough: How Constraints Beat Brute Force

DeepSeek entered the stage when the industry giants were building ever-larger models, training them on colossal datasets, and running them on massive infrastructure. This was an era dominated by brute force, in which scale itself was a key indicator of progress.

The AI giants were like heavyweight boxers relying on computing power to knock out every problem. DeepSeek entered the ring like a mixed martial artist who uses leverage, efficiency, and technique to win, demonstrating that intelligence and agility can beat sheer force.

Cracks were appearing in the strategy of building large models. Training costs were ballooning. Latency and operational expenses remained stubbornly high. More troubling, the incremental gains from adding more parameters were diminishing. The brute-force approach was delivering progress, but at a cost that was becoming harder to justify.

Although some industry players explored model efficiency, a systematic approach that truly balanced scale with architectural innovation was an open opportunity.

© Harshad Oak, Monish Darda 2026
H. Oak and M. Darda, *The DeepSeek Moment*, https://doi.org/10.1007/979-8-8688-2598-9_2

Into this landscape, DeepSeek introduced a different philosophy. Instead of merely scaling up, it asked a sharper question: what if efficiency itself was the path to intelligence? This view did not deny the value of scale but balanced it with efficiency. DeepSeek made careful trade-offs in compute, memory, and algorithmic design, demonstrating that the game was not merely about who had better hardware and deeper pockets (see Table 2-1).

Table 2-1. *The Rules of the Game Have Changed*

Old Era (Pre-DeepSeek)	New Era (Post-DeepSeek)
"Bigger is better."	"Smarter is better."
Benchmark worship	ROI (return on investment) and TCO (total cost of ownership) focus
Closed, elite labs	Open, distributed innovation
Data hoarding	Model efficiency and reuse
Capital moat	Design moat

Constraints as Catalysts for Innovation

Throughout the history of technology, the most powerful innovations have often been sparked not by unlimited resources, but by tight, unforgiving constraints. When brute force isn't an option, ingenuity is forced to take over, and the results can change the world.

In the early days of personal computing, IBM PCs were limited to just 640 kilobytes of usable memory, a tiny fraction of the size of a single photo today. However, that severe limitation forced a generation of programmers to invent clever memory management tricks to do more with less.

This ethos of squeezing performance from scarcity wasn't just confined to software. The smartphone revolution, for instance, would not have been possible with power-hungry desktop processors. The absolute constraint of battery life led to the rise of ARM processors. The ARM design wasn't the most powerful, but it was perfect for the job. It now powers virtually every phone on the planet and has made significant inroads into the laptop market through Apple Silicon M-series chips and Qualcomm's Snapdragon processors. Most of this book is written using laptops powered by the ARM processor!

Some of us may remember the iPod and Napster. That wave of transformation was fueled by the MP3 file format, an innovation born from the challenge of sending music over slow phone lines without distortion. Initially intended for broadcasters, the format spread to the internet and sparked a revolution that upended the entire music industry.

From the memory limit in a PC, to the battery in your pocket, and to the sound of a human voice, the pattern is the same. A hard limitation forced a more innovative, more elegant solution than brute force could ever provide. It's a lesson that has reshaped entire industries, and DeepSeek resurrected this ethos for the age of large-scale AI.

The constraints were real. A frontier model could cost hundreds of millions to train. Power consumption was measured in gigawatt-hours, and the carbon footprint of scaling drew increasing scrutiny. For many firms, the costs were simply too high.

DeepSeek responded by treating these limits not as weaknesses but as design prompts. Its engineers asked fundamental questions that flipped the "more is better" assumption:

- If only a fraction of the model is needed for a request, why compute all of it?

- If memory is a bottleneck, why not reduce precision in ways that save cost while preserving accuracy?

- If datasets contain substantial redundancy, why not curate them better to avoid wasted training cycles?

Each question reframed the goal from maximizing outcomes to improving outcomes per unit of resource.

This reorientation also shifted how success was measured. Rather than chasing the largest parameter count, DeepSeek benchmarked progress in cost-per-query, latency under load, and energy consumption. These metrics resonated not only with researchers but also with enterprises and policymakers who prioritized accessibility and sustainability as much as raw power.

In short, DeepSeek proved that innovation thrives under constraint. The following sections unpack the specific strategies that enabled this efficiency-first philosophy to succeed where brute-force scaling was reaching its limits.

The Starting Point: How LLMs Work

To understand DeepSeek's architectural breakthroughs, it is helpful to first examine the foundation on which it was built: the standard Transformer-based large language model (LLM).

At its core, an LLM is a prediction machine. Think of the autocomplete feature on your phone, but scaled to an unimaginable degree. It tries to guess the next word, or, more precisely, the next "token," based on the ones that came before it. Tokens are the basic building blocks of text; they can be whole words, but are often just pieces, like pre- or -ing. When trained on trillions of words, this simple guessing game scales into something extraordinary.

The architecture that makes this possible is the Transformer, introduced in the now-famous 2017 paper, "Attention Is All You Need." The key mechanism is self-attention, which enables the model to weigh the relationships between words no matter how far apart they are in the text. For example, in the sentence, "The trophy would not fit in the suitcase because it was too small," self-attention enables the model to correctly determine that "it" refers to the suitcase and not the trophy by considering all contextual relationships simultaneously.

The trade-off for this powerful contextual awareness, however, is immense cost. The computational cost of self-attention scales quadratically. In simple terms, doubling the length of your text doesn't just double the work. It quadruples it.

To use a gaming analogy, it's like adding a new player to a massive online multiplayer game. The server doesn't just track the new player; it must constantly calculate their line of sight, distance, and potential interactions with every other player on the map. With each new person who joins, the server's workload multiplies. This computational bottleneck is what makes running large-scale models so incredibly expensive.

Training involves exposing the model to vast amounts of text and gradually adjusting its billions of internal parameters until it becomes highly effective at predicting the next token. Some experts view this process as an advanced form of compression. The model effectively digests a library's worth of books and stores the essence of that knowledge—the concepts, patterns, and relationships within its parameters.

However, this is fundamentally different from a simple compression tool like a ZIP file. ZIP files store data perfectly for exact reconstruction (lossless compression). An LLM doesn't just store the text; it learns the rules that govern it. This is why it can't reproduce a book word for word, but it can write a new paragraph in the author's style. It is from this generative quality that the surprising, complex abilities emerge: the capacity to summarize books, write functioning code, and even reason through problems its designers never explicitly programmed.

This discovery, that massive scale leads to these abilities, was the spark that ignited the brute-force era. The path to more powerful AI seemed to be a straightforward, if colossally expensive, race. Feed the machine more data, train and run it on more powerful hardware with more parameters and more context, and it would get smarter. This is the foundational assumption that DeepSeek challenged.

DeepSeek's Architectural Edge: Rethinking AI Foundations

DeepSeek's rise was not merely about resource discipline; it was about reengineering LLM architectures to improve efficiency. Its edge came from several intertwined innovations that delivered compounded gains.

Using a Mixture-of-Experts (MoE) for Efficiency

A key innovation was the use of a Mixture-of-Experts (MoE) architecture. Think of it like booking a ride on Uber. When you request a car, the app does not alert every single driver in the city. Instead, it instantly matches you with the drivers best suited to pick you up based on location and availability. That way, the system is fast, efficient, and scalable. MoE works similarly. Instead of activating the entire model for every task, a small router network directs the request only to the experts best equipped to handle it.

This approach was not just a theory. DeepSeek's MoE models delivered performance on par with much larger dense models but with far less training compute. For example, DeepSeek-V2 has 236 billion parameters, making it highly capable; however, for any given task, it activates only about 21 billion per token, less than 10% of the total. DeepSeek-R1, even larger at 671 billion parameters, activates only about 37 billion per token.[1] In practice, this means the models run with the efficiency of a smaller system while keeping the overall capacity of a much bigger one.[2]

At inference time, each query activated only about 10% of the parameters, significantly lowering the cost-per-query compared to its peers.

---

[1] https://arxiv.org/pdf/2501.12948

[2] Dense Training, Sparse Inference: Rethinking Training of Mixture-of-Experts Language Models, arXiv https://arxiv.org/html/2404.05567v1

It is essential to clear up a common misunderstanding about what "experts" in MoE actually are. They are not subject matter experts like medicine or law. Instead, they are smaller sub-networks within the model that process data in slightly different ways. Their specialization emerges statistically during training, not by design. Some may become better at handling structured sequences such as code, while others lean toward more natural or figurative text. These are not rigid roles, but they allow the system to distribute work effectively.

When the model receives a prompt, the router analyzes the text and decides which experts are most likely to process the next token. Only a few are selected from many. This dynamic selection is what makes MoE efficient. Instead of engaging the entire network at every step, only a fraction of its capacity is active at a time, keeping computational cost and memory footprint manageable while still enabling scaling to hundreds of billions of parameters.

Efficiency matters not only for speed and cost but also for environmental considerations. Training a model consumes vast amounts of energy. Still, over its lifetime, the ongoing use of the model, inference where millions of users query it daily, often has an even larger environmental footprint. Highly efficient models, such as DeepSeek, significantly reduce inference costs, thereby lowering energy consumption and increasing cumulative ecological savings.

Making the AI's "Working Memory" More Efficient

Every conversation you have with an AI comes with a hidden cost. To track the dialogue, the model maintains a running record in its key–value cache, a form of working memory. As the chat gets longer, this memory swells, forcing the AI to work harder just to keep up. It's a bit like trying to carry an entire library on your back just to find the next page you need.

The engineers at DeepSeek saw this as an opportunity. Their innovation, Multi-head Latent Attention (MLA), reimagined how an AI's memory could work.

Instead of storing the entire chat history in full detail, MLA selectively compresses older parts of the conversation into a smaller latent space. This creates a mathematical shorthand that preserves the essential meaning while reducing the data burden. Think of it like keeping the current page of a book in front of you, while turning all the previous chapters into a crisp, one-page summary.

This simple change had a huge impact. By reducing the memory footprint, MLA reduced expensive data transfers and freed hardware to support many more users simultaneously. The result was faster responses at a lower cost.

Unlocking Long Contexts with Sparse Attention

When a standard AI analyzes a lengthy document, it exhaustively compares every word (or token) with every other word, a slow and computationally expensive process. Newer Deepseek models use Sparse Attention, a method that intelligently prunes or trims away the redundant and unimportant connections between words.[3] It's trained to identify and preserve only the most critical dependencies that determine the document's meaning, much like an expert analyst skims a report for key data points. This allows the AI to process very long documents many times faster, unlocking new capabilities in fields that rely on long-form content, such as legal contract review, financial analysis, and scientific research.

---

[3] `https://github.com/deepseek-ai/DeepSeek-V3.2-Exp`

Using a More Resource-Friendly Numerical Language

DeepSeek applies a technique called quantization, which means it uses smaller, lighter kinds of numbers for its calculations. Typically, AI models rely on detailed numbers, 16-bit or 32-bit values, to track probabilities and weights inside the network. These are like carrying around measurements down to many decimal places.

With quantization, DeepSeek switches to simpler formats:

- **FP8 (Floating Point 8-Bit)**: A compact way of representing numbers that still allows decimals but with fewer digits of precision

- **INT8 (Integer 8-Bit)**: An even simpler format that only stores whole numbers in a limited range

Both are lighter than FP16 or FP32, meaning they use fewer computer bits to store each number. DeepSeek trained its models from the ground up to work with these formats, so the loss of precision is minimal.

The benefit is significant: quantization reduces the memory needed to store data and the energy required for calculations. Industry benchmarks indicate savings of approximately 40% relative to higher-precision formats.[4] This makes DeepSeek's models faster, less computationally expensive to run, and more practical to deploy on widely available hardware.

Hardware–Software Co-design

One of DeepSeek's most important lessons is that actual efficiency comes not from more innovative algorithms or more powerful chips alone, but from designing them to work together in perfect sync. This principle, called hardware–software co-design, is like the difference between buying

---

[4] https://arxiv.org/html/2411.06084v1

a suit off the rack and having one custom-tailored. While the raw materials may be the same, the bespoke suit fits perfectly, providing comfort and performance that a generic one never could.

Instead of building a model and simply running it on available hardware, DeepSeek's engineers ensured their software and hardware were in constant conversation. Their model architecture was specifically tailored to exploit the strengths of the chips it ran on, like Nvidia's H800 accelerators.

This tailoring involved several key decisions:

- **Smarter Expert Routing**: In a Mixture-of-Experts (MoE) model, data is sent to "specialist" parts of the network. Co-design ensures these specialists are placed on the GPU most logically, like organizing a hospital so the cardiac surgeon's office is right next to the cardiac imaging lab. This minimizes the "travel time" for data, reducing delays.

- **Efficient Memory Management**: This is like designing a custom warehouse system. Instead of placing items randomly, the system knows which tools are used together and stores them on the same shelf. This enables the GPU to fetch the required data more quickly, thereby increasing overall performance.

- **Optimized Numerical Precision**: The model was tuned to use simpler number formats (FP8 quantization). This is like choosing to save an image as a highly efficient JPEG instead of a massive RAW file. For most tasks, the quality is indistinguishable, but the file is orders of magnitude smaller and is processed faster.

The business impact of this custom tailoring is direct and significant. Co-design delivered much higher throughput per dollar than running a standard Transformer on generic GPUs. For applications in document-heavy fields like contract analysis, this translated into noticeably faster response times, bringing AI out of the lab and into production use.

The Business of Efficiency

DeepSeek's technical foundation translated directly into business impact. The effect on the market was similar to what happened with cloud computing a decade earlier.

Before cloud services like AWS and Azure, startups simply couldn't compete with the giants, who could build mega data centers requiring massive, upfront capital. Cloud computing leveled the playing field by offering instant, pay-as-you-go access to world-class storage and computing. Suddenly, a small team could build and scale a global application without owning a single server.

DeepSeek's efficiency did something similar for AI. By drastically lowering the cost and hardware needed to run powerful models, it gave smaller companies access to capabilities that were previously the exclusive domain of the tech giants, much like how the cloud wave enabled the rise of companies like Icertis. In the same way, DeepSeek is likely birthing a new generation of unicorns and business models built on accessible AI.

Through 2025, DeepSeek's share of the enterprise AI market was increasing, particularly in cost-sensitive sectors such as finance and healthcare. These organizations weren't just chasing benchmark scores; they were looking for a practical solution that was affordable enough to deploy widely. We will explore in detail in later chapters.

Redefining the Possible

DeepSeek's breakthrough was not about rejecting scale but about making it smarter. The industry's prevailing logic was that progress demanded more: more data, more compute, more parameters. DeepSeek proved that progress could also come from less, if applied with architectural wisdom. It also showed what a small team of talented people could achieve, not despite their limitations, but because of them.

Scale remains essential for pushing the boundaries of AI, and DeepSeek's models are themselves massive. But their success showed that scale without efficiency is a wasteful and exclusionary strategy. The future belongs not just to those who can build the most extensive systems, but to those who can build the most intelligent. DeepSeek proved that in the new era of AI, efficiency is no longer a compromise, but a significant competitive advantage.

CHAPTER 3

Beyond the Leaderboard: When Efficiency Redefined Winning

After a couple of years of generative AI dominating the public mind, the industry's pecking order had been set. Progress was measured by incremental gains on benchmarks, with each new model version reinforcing the "bigger is better" doctrine examined in earlier chapters.

Then DeepSeek showed up at the contest. At first, rivals and analysts treated it as an intriguing outsider. Clever architecture, some cost-saving tricks, perhaps a sign of interesting research from China. But when its models, built on the efficiency-first design we explored in the last chapter, went head-to-head with GPT, Claude, Gemini, and Llama, the conversation changed. In reasoning tests and real-world tasks, DeepSeek's models were standing shoulder to shoulder with those from the AI giants.

This was the inflection point. When David met Goliath, the remarkable fact was not only that David landed a strike. It was that the crowd watching the duel suddenly began to question the rules of the game.

© Harshad Oak, Monish Darda 2026

H. Oak and M. Darda, *The DeepSeek Moment*, https://doi.org/10.1007/979-8-8688-2598-9_3

When global investors dumped tech stocks in response to DeepSeek, the shock of performance matching was validated by the most ruthless benchmark of all: the markets. This exposed a fundamental flaw in the industry's obsession with its own scoreboard.

When Benchmarks Stopped Being Enough

This shockwave forced a new way of thinking. If one model from China, trained at a fraction of the perceived cost at a little-known startup, could rival America's best, what did these benchmark numbers really mean?

Enterprises began to ask more practical questions:

- How much does it *cost* to run this model (the cost of inference)?

- How many GPUs does it consume?

- How predictable is its performance in production?

- Is it safe to use?

It was as if the contest had shifted from building a supersonic fighter jet to building a lightweight drone that was stealthier, got the job done, and did it at a fraction of the cost and fuel consumption.

This forced a shift in enterprise priorities. In boardrooms worldwide, the language around AI began to change. A year earlier, executives were interested in *the best-performing model on specific benchmarks or tasks, the state-of-the-art (SOTA) model.* Now they were asking: *which model gives us the best return on investment?*

DeepSeek's model family provided a natural answer. R1 was the reasoning specialist—slow, meticulous, and often more expensive per query, but great at complex problem-solving. V3, by contrast, was fast, versatile, and efficient across writing, summarization, and simpler code. Both were substantially cheaper than their closest competitors.

Redefining the Scoreboard

The shock of exceeding performance and price benchmarks by magnitudes did more than just introduce a new competitor. It forced the industry to admit that its primary tools of measurement, the benchmarks that defined "SOTA," were flawed and incomplete. To understand the shift, we must first understand the scoreboard everyone had been using.

The Old Scoreboard: An Exam for Mimics

For years, the industry was locked in a "leaderboard addiction." The belief was that if a model got a high score on a standardized test, it must be getting smarter.

It was like a standardized aptitude test for machines.

The tests became famous in their own right. MMLU (Massive Multitask Language Understanding) was the "general knowledge" section, drawing on thousands of multiple-choice questions across 57 subjects, including law, history, and physics. GSM8k (Grade School Math) was the "logical reasoning" section, comprising word problems. HumanEval was the "essay" portion, asking the AI to write simple, self-contained blocks of code.

To understand the industry's mindset, it helps to see what these exam questions look like:

- **The Knowledge Test (MMLU):** "A 33-year-old man undergoes a radical thyroidectomy. Postoperatively, serum calcium is 7.5 mg/dL. Damage to which of the following vessels caused the findings?"

- **The Reasoning Test (GSM8k):** "Janet's ducks lay 16 eggs per day. She eats 3 for breakfast and sells 9 eggs a day at the bake sale. Each batch of muffins requires 4 eggs. How many batches of muffins does she bake each day?"

- **The Coding Test (HumanEval):** "Create a function called has_close_elements that checks if any two numbers in a list are closer to each other than a given threshold."

This approach, however, created two major problems. First, many felt that we were teaching models to pass standardized tests of intelligence, not to live life. A model could score 90% on a test but be totally unable to help you plan a business trip.

Second, the tests were vulnerable to being gamed. These benchmarks were static, and their questions were widely available online. As labs trained new models on the entire web, they were *unknowingly feeding them the answer key*. A high score no longer guaranteed the model was *reasoning*; it might just be *remembering*.

The New Guard: From Scholar to Surgeon

By 2025, a critical realization emerged: an AI's ability to pass a static exam was no longer a sufficient indicator of its worth. The enterprise imperative shifted. The market no longer valued a mere scholar; it demanded a surgeon, a specialist who could apply that knowledge within a complex, high-stakes system and perform the necessary operation.

This practical need gave rise to a wave of agentic benchmarks, tests designed to measure an AI's ability to act as an autonomous agent, not just answer questions. These were not exams; they were obstacle courses:

- **SWE-Bench:** This was the ultimate "surgeon's test." It did not ask the AI to write a simple function. It pointed the AI to a real bug in a massive, open source project and said, "Fix this." The AI had to read the complex, human-written codebase, diagnose the problem, and write a patch that actually worked without breaking anything else.

- **Terminal-Bench**: This tested if the AI could operate a computer terminal—that black-and-white command line. It was like watching an expert user try to install software or manage files, all through text commands.

This new, practical scoreboard became the focus of all the top AI labs. Model releases from DeepSeek, Google, and Anthropic in 2025 no longer led with MMLU scores. They now prominently featured their performance on agentic tests like SWE-Bench. The industry-wide narrative had shifted from the "scholar's exam" to the "surgeon's test."

The results on this new scoreboard showed a tight, competitive race. On any given day, a different model might pull ahead on a specific agentic benchmark.

This achievement was not just about "smarter" AI; it was about more useful AI. And it led directly to the final, unavoidable metric.

The Final Metric: Efficiency

Yet even these modern benchmarks carried a hidden flaw: they still ignored cost. An AI could top every agentic chart but be so slow and expensive that no business could afford to run it.

This is where DeepSeek's efficiency-first philosophy made its case undeniable. The conversation shifted because DeepSeek had *already* achieved high scores on the old benchmarks. This created a crisis for the incumbents.

If Model A (GPT-4) and Model B (DeepSeek) both get a 90% on MMLU, but Model B costs one-tenth as much, is the score really the thing that matters?

The answer was a resounding "no." The *cost to achieve the score* became the new, more important metric.

What had been a niche academic discussion under the banner of "Green AI" suddenly became a core business concern. The new, multidimensional scoreboard now included

- **Performance-Per-Dollar**: How much intelligence do I get for my money?

- **Performance-Per-Watt**: What is the energy and environmental cost of this query?

- **Total Cost of Ownership (TCO)**: What is the *all-in* cost to run this model on my own hardware, including inference, energy, and engineering time?

In this new game, efficiency was not just a feature; it was a differentiator. A model that was "90% as good at one-tenth the cost" was no longer a compromise. For most businesses, it was a brilliant strategic victory.

How to Read the New Scoreboard: The "Spec Sheet" vs. the "Arena"

This new, multidimensional reality left many leaders with a simple question: "If the old scoreboard is broken, how do I actually *choose* a model today?"

As of late 2025, the industry has settled on two ways of measurement. To make a smart decision, you must understand both.

The "Spec Sheet" (Quantitative Leaderboards)

The first way to measure a model is the "spec sheet." This is the approach used by sites that aggregate dozens of benchmarks. It is a comprehensive quantitative dashboard listing scores for everything from MMLU and GSM8k to newer agentic tests like SWE-Bench.

- **The Analogy**: This is a car's technical specification sheet. It gives you the raw, objective numbers in a controlled lab: 0-60 time, horsepower, torque, and braking distance.

- **The Value**: This is what your engineers need. It tells you the absolute technical limits of a model. Can it handle high-precision math? Can it fix complex bugs? Can it go up a 45-degree incline?

- **The Flaw**: It tells you nothing about *how the car feels to drive*. It is on-paper performance.

The "Arena" (Human-Preference Leaderboards)

The second way to measure is the "test drive." This is the philosophy behind platforms such as the Arena, which has become a notable industry benchmark.

The concept is simple: thousands of real users are served two anonymous model answers to the same prompt. They vote for "which answer felt better." There is no "right" answer, only user preference.

- **The Analogy**: This is the *test drive*. You get behind the wheel and see how it feels on a real road.

- **The Value**: The Arena measures the "vibe." It's a qualitative score for hard-to-define qualities like helpfulness, common sense, and even tone. It tells you how the model *feels* to work with.

- **The Flaw**: It can be swayed by style over substance. A "nicer" or more verbose AI might win votes over one that is more accurate but blunt.

The Leader's Takeaway

To make a smart decision, you must look at both. The spec sheet tells you what a model *can* do. The Arena tells you what it *feels* like to use.

This dual-measurement system perfectly explains the two strategic decisions we'll explore next. The "business sense" choice was an Arena-driven decision, while the "mission sense" choice was a spec sheet–driven one.

From SOTA to ROI: Two Decisions

This new, nuanced view of "best" played out in executive suites and government offices.[1] The choice of an AI model was no longer a single decision but a strategic trade-off.

Case 1: The "Business Sense" Model

At one company, the CIO had been pushing for a full-scale GPT integration for their global customer service platform. The model was powerful, but the projected monthly cloud inference bill ran into millions of dollars.

During the budget meeting, a younger engineering manager spoke up: "Why not test DeepSeek-V3? It's not just open source, it's MIT-licensed. We can deploy it on our own hardware. Even if accuracy is slightly lower, the savings are massive."

The finance chief perked up immediately. *"How massive?"*

The engineering manager laid out the numbers: a projected 40% cost reduction for the same volume of customer interactions, with negligible differences in customer satisfaction scores. The room went quiet. Then the CEO asked the only question that really mattered: "If our customers do not notice the difference, why should we?"

---

[1] Illustrative examples based on industry experiences.

That single exchange captured the essence of the shift. Benchmarks might have crowned some model as the best, but inside enterprises, the crown was being passed to whichever model delivered the best return on investment.

Case 2: The "Mission Sense" Model

But not every organization made the same choice. A research agency evaluating large language models for analysis reached a different conclusion.

Their use case required mathematical precision and logical reasoning over long chains of evidence. These were tasks in which even small inaccuracies could have large consequences. The team ran side-by-side tests of DeepSeek-R1 and DeepSeek-V3.

V3 was faster and cheaper, but R1 consistently outperformed it in structured problem-solving and reasoning depth. The difference was enough to shift the risk calculus. The agency chose R1, accepting the higher compute costs as the price of precision.

One of the researchers summarized it neatly in their internal memo: "V3 makes business sense. R1 makes mission sense."

DeepSeek's dual-model strategy, offering one line optimized for cost and another for intelligence, gave every type of buyer something to call "best."

The Ripple Effect: Efficiency As Liberation

The real proof came from the developer community. Within weeks of release, DeepSeek's models climbed the Hugging Face download charts. Startups, eager to avoid the steep licensing fees of proprietary systems, began exploring DeepSeek for their products.

The MIT License, first adopted by DeepSeek starting with DeepSeek-R1 worked as a catalyst. It meant anyone could deploy DeepSeek commercially, free of royalties or hidden clauses. For an industry weary of restrictive "open-but-not-really" licenses, it felt like liberation.

When the cost of experimentation falls, creativity rises. A fintech startup in Bangalore reported drastically cutting its chatbot operating costs. An ecommerce firm in Jakarta fine-tuned V3 on local data to get the desired performance at a fraction of the cost of using a proprietary API.[2]

These were not headline benchmarks. They were quiet, practical validations. Analysts began calling it the ROI Cascade, the economic equivalent of a viral contagion. Once one enterprise demonstrated that a lower-cost model could deliver comparable results, others were compelled to follow.

AI had been a luxury market. DeepSeek turned it into a utility market.

Victory Over the Scoreboard

Looking back, the moment of performance matching was only the surface-level story. The deeper impact was that it forced the industry to measure, think, and buy differently.

DeepSeek's lasting contribution was not simply proving that efficiency could match scale. It was proving that efficiency could redefine victory. In that sense, its most significant triumph was not over any single competitor, but over the industry's old scoreboard itself. There will be better and more efficient models in the future—the DeepSeek moment will always be remembered as the one that was the inflection point of getting ROI front and center in the application and evolution of generative AI.

---

[2] Illustrative examples based on industry experiences.

CHAPTER 4

Leveling the Playing Field: The Post-DeepSeek Democratization

The End of the High-Cost Era

For millennia before Johannes Gutenberg invented the printing press (~1,440 CE), knowledge was a proprietary asset. It was capital-intensive, existing only in hand-scribed manuscripts locked away in exclusive institutions and private libraries. It was controlled by a tiny elite who possessed the resources to create it and had the specialized training to consume it. The printing press was a democratizing force, not because it made books better, but because it reduced the cost of replication to near zero. It broke the information monopoly and transferred power from the few to the many.

In the early 2020s, artificial intelligence was in its own pre-Gutenberg era. It was a capital-intensive asset, locked away in the walled gardens of a few technology giants. The cost to train a frontier model was measured in

© Harshad Oak, Monish Darda 2026
H. Oak and M. Darda, *The DeepSeek Moment*, https://doi.org/10.1007/979-8-8688-2598-9_4

the hundreds of millions of dollars, requiring data centers that consumed as much electricity as small cities. This created a tangible fear across the business world that a handful of trillion-dollar corporations would dictate the future of human intelligence.

This monopoly was particularly jarring because it represented a reversal of the founding promise of the industry's frontrunner. OpenAI was founded in 2015 as a nonprofit organization with a clear mission to benefit humanity as a whole, unconstrained by the need to generate financial returns. The core belief was that AI should be distributed as broadly and evenly as possible. However, in 2019, OpenAI pivoted. It established a capped-profit subsidiary to absorb billions in investment from Microsoft and others, followed by the closed-API release of its most powerful models. It was seen by the community as a philosophical rupture, with one media report stating that OpenAI was "now everything it promised not to be: corporate, closed-source, and for-profit." The champion of Open had become Closed, leaving an ideological vacuum in the market.

The open source movement, chaotic but resilient, rushed to fill it. The first signs appeared in March 2023, not through a product launch, but through a leak. Meta had announced its Llama model as a research-only tool, licensed carefully to select academics. Days later, the full model weights were leaked online via a torrent on 4chan. A powerful foundation model was now available far beyond its intended audience.

The leak provided the blueprint, but the Vicuna project provided the price tag. Almost immediately, researchers from UC Berkeley and Stanford took the leaked Llama model and fine-tuned it. Their method was ingenious. They scraped 70,000 user conversations from ShareGPT, a site where users shared their ChatGPT outputs, and used the outputs of OpenAI's closed model to train the open source competitor.

The shock was not that they did it, but that they did it so cheaply. The total training cost for Vicuna was estimated at approximately $300.

It forced every CTO and investor to ask a question: Is the $100 million capital moat actually a myth? If a university team could replicate 90% of the capability for the cost of a weekend trip, was the era of the AI Walled Garden ending, and was "good enough" AI now effectively free?

Open Source vs. Open Weight: The War of the Fine Print

The leak demonstrated that democratization was possible, but the enterprise world required it to be legal. You cannot build a Fortune 500 AI strategy on a model found on a message board. Meta responded in July 2023 by officially releasing Llama 2 for commercial use. It was a watershed moment, but it came with asterisks. The license included restrictions on the number of users. If you had more than 700 million monthly active users, you must request a license from Meta.

This is where the DeepSeek moment of January 2025 became catalytic. DeepSeek did not just solve the technical efficiency problem. They also resolved the rights issue. By releasing DeepSeek-R1 under the permissive MIT license (with V3 following in March 2025), they removed the barrier to entry. There were no user caps, no commercial restrictions, and no fine print preventing a competitor from using the model. This decision sparked the Cambrian Explosion, a sudden burst of diverse new life forms in the AI ecosystem.

To understand why this distinction matters so much to business leaders, we must distinguish between closed source, open source, and open weight. You can think of AI models not as software, but as food.

Closed Source Is Fine Dining

Models like GPT are like an exclusive, reservation-only restaurant. You pay for a world-class meal, and it is delicious. But you must eat it on their premises. You cannot see the kitchen, you do not know the ingredients or the secret spice blend, and you certainly cannot take the recipe home to cook it yourself. You are a consumer, dependent entirely on the restaurant.

True Open Source Is Grandma's Cookbook

This is the Linux model. You inherit the master chef's handwritten cookbook. You get everything, including the exact list of raw ingredients or data, the exact techniques or training code, and the step-by-step process. You have the full recipe to cook the meal yourself from scratch in your own kitchen, modifying it however you see fit.

Open Weight Is Gourmet Takeout

This is the model adopted by DeepSeek and Llama. You do not get the secret recipe. The raw training data is not disclosed. But crucially, you own the food. You can take the dish home. You can serve it in your own kitchen. You can analyze it in a lab. You can add your own spices to change the flavor profile. You have total control over how, when, and where the meal is served.

The industry rallied around open weight because it offered the convenience of a pre-trained brain without the lock-in of a closed API.

This brings us to Distillation, a key technique in the new efficiency playbook. Following our culinary analogy, if the large model is the Master Chef, a distilled model is their talented Apprentice. The Master Chef teaches the Apprentice by having them taste and recreate thousands of complex dishes. The Apprentice lacks the Chef's 40 years of wisdom, but they learn to cook 90% of the menu just as well and work much faster and for a lower salary.

Many of the new open-weight models are built on this principle. DeepSeek was not an anomaly. It was the vanguard of a new, hyper-competitive open-weight AI.

Why Give Away the Crown Jewels?

Why would Meta and DeepSeek give away assets worth hundreds of millions of dollars for free? It was not charity but a calculated strategy against established competitors.

Meta Is Pursuing a Strategy to Commoditize the Complement

If you can make a complementary product cheap or for free, the demand for your core product goes up. Google is the modern master of this strategy. They released Android and Chrome for free, effectively driving the price of mobile operating systems and web browsers to zero. Google did not care about selling software licenses. They cared that a cheaper, faster, ubiquitous web meant more people searching and clicking Google Ads.

For Meta, the AI model is the complement. Their core business is social media and advertising. Their rivals, OpenAI and Google, sell access to models as a product. By releasing Llama for free, Meta effectively drives the price of intelligence down to zero. This reduces OpenAI's and Google's profit margins while preventing any single rival from becoming a gatekeeper that could charge Meta a tax for using AI in the future.

DeepSeek Is Pursuing the Challenger's Gambit

Why would a for-profit entity give away an asset so valuable? It was a calculated strategic move designed to destabilize the incumbents.

First, it solved the trust barrier. Western enterprises are skeptical of sending private data to a closed API hosted in China. By open sourcing the weights, DeepSeek allowed companies to run the model on their own servers, behind their own firewalls. You do not have to trust DeepSeek with your data. You can inspect the model and run it yourself.

Second, it was a scorched-earth strategy. By demonstrating that they could train a frontier-class model at a fraction of competitors' costs, DeepSeek signaled to the market that Western AI prices were artificially high. This forced a pricing correction across the industry that hurt the high-margin incumbents far more than it hurt the challenger.

The Broader Pattern of Democratization

The DeepSeek moment, while shocking in its speed, is not an anomaly. It is the next logical step in a recurring historical cycle of technology democratization. For business leaders, this arc should be recognizable.

Consider the Personal Computer Revolution. In the 1960s and 1970s, computing power lived in climate-controlled rooms guarded by specialists in white coats. Access was rationed. Mainframes were expensive to purchase and to operate. When the PC arrived, it did not match the power of the mainframe, but matching the mainframe was not the point. It put computation on every desk in every home and office.

Cloud computing was a similar paradigm shift. Before AWS, a startup needed to buy physical servers and rent data center space. This massive Capital Expenditure (CapEx) was a barrier to entry. The cloud turned that CapEx into a variable Operational Expenditure. Suddenly, a two-person bootstrapped startup could rent the same infrastructure as a Fortune 500 company.

The parallel is exact. The giants of Chapter 1 built the Big AI mainframes. DeepSeek is the PC.

Developers can now even run DeepSeek-R1 derivatives and distilled versions locally on their laptops, completely offline. It is not just that servers are cheaper. For a class of use cases, servers are no longer necessary. This is the ultimate democratization. High-level intelligence running on consumer hardware, independent of any cloud provider.

DeepSeek and Llama have democratized the intelligence layer just as Linux democratized the operating system and cloud computing democratized the server.

The New Strategic Calculus: Rent vs. Own

This shift fundamentally changes the decision matrix for enterprise leaders. For the past three years, the default strategy has been rent. You paid OpenAI or Google for access to their API. Now, you have a choice.

Renting remains the appropriate choice for many. It offers speed and simplicity. You do not need an engineering team to manage infrastructure. You submit a query and receive an answer. It also provides access to the absolute peak of capability. State-of-the-art models often retain a slight edge over open models in pure reasoning power. For rapid prototyping or non-core functions, renting is efficient.

However, for the core business, owning has become the superior strategy for three critical reasons.

The first is the data moat. In a world where the model itself is a commodity available to everyone for free, the only durable competitive advantage is the proprietary data used to fine-tune it. If you rent a generic model, you get generic intelligence. If you own an open model, you can inject your unique institutional knowledge, such as your customer logs, your proprietary codebases, and your patient records, into the model's very DNA. This creates a specialized asset that no competitor can replicate, because they lack your data.

The second is control and privacy. This is the primary driver for regulated industries. When you own the model, it runs inside your firewall. Your data never leaves your perimeter. As the Samsung incident demonstrated, in which confidential code was leaked to ChatGPT,[1] this control may be non-negotiable for certain use cases in domains such as finance, healthcare, and defense.

The third is cost predictability. Renting is cheap to start, but it becomes punishingly expensive at scale. Owning shifts the cost from a volatile per-token tax to a predictable fixed cost for hardware and operations.

The New AI Economy Takes Shape

Democratization did not just lower costs. It birthed an entirely new economy of tools and talent.

The revolution needed a central hub, and Hugging Face cemented its role as the GitHub of AI. It became the indispensable distribution layer for the movement. By late 2025, DeepSeek-R1 derivatives had amassed millions of downloads on the platform. It provided the social and technical fabric that enabled this rapid innovation, hosting the community that fine-tuned and turned raw materials into usable products.

Alongside the tools came a permanent shift in talent. We witnessed the maturation of the AI engineer. In the early days, building a viable AI product was a scientific endeavor requiring PhDs to train models from scratch. The advent of APIs and early open models such as Llama began to change this, but DeepSeek removed the final barrier to entry. Today, the AI engineer does not need to know how to train a model; they only need to know how to orchestrate one. By treating open-weight models as modular, free components, they can integrate code and intelligence to build complex applications. DeepSeek fueled this democratization, ensuring that a two-person startup with some tech skills could build systems that previously required mega research budgets.

---

[1] https://www.forbes.com/sites/siladityaray/2023/05/02/samsung-bans-chatgpt-and-other-chatbots-for-employees-after-sensitive-code-leak/

The final validation of this new economy came when the giants themselves embraced it. We saw a replay of the Red Hat strategy from the Linux era. Just as Red Hat built a billion-dollar empire not by selling free software, but by selling the stability and security to run it, companies like Red Hat and others emerged to provide the enterprise-grade scaffolding for free models. Even the cloud hyperscalers offered DeepSeek as a fully managed service on their cloud. They realized that in the new economy, they could not just sell their own models; they had to be the platform for all models.

The definitive signal that the AI world had changed came on August 5, 2025. For years, OpenAI had been a defender of closed models, arguing that open source models were dangerous and commercially unviable. But the pincer movement of Meta's Llama and DeepSeek's efficiency left them little choice.

With an eye on the enterprise on-prem market, OpenAI changed course. They released gpt-oss-120b, their first open-weight model since GPT-2 in 2019. Sam Altman, in a US Senate hearing in May 2025, said,

> *"At OpenAI, we're committed to the path of democratic AI..."*
> *"We want to open source very capable models."*[2]

Open source GPT was released soon after.

It was a tacit admission that in a post-DeepSeek world, you cannot win by renting intelligence alone. You must offer the takeout option, or the market will go to the chef who does.

Distributed Power, Distributed Risk

The democratization of AI is not a utopia. When a company rents intelligence from a closed API, it is also renting that vendor's massive trust and safety team. When an enterprise owns an open model, it also

---

[2] `https://www.commerce.senate.gov/wp-content/uploads/media/doc/Sam%20Altman%20-%205-8-25%20Senate%20Commerce%20Committee%20Testimony.pdf`

inherits the responsibility. The burden of security, governance, and ethical compliance shifts from the API provider to the enterprise.

This risk is not theoretical. The new, open ecosystem has already provided stark proof of the dangers. Early audits of DeepSeek derivatives revealed high jailbreak rates, in which the models could be tricked into bypassing their safety filters.

For the enterprise, this responsibility transfer is the primary source of friction. However, it also creates the next great business opportunity in selling trust, security, and compliance as a service.

The Next Battleground

For leaders, the strategic playbook has been rewritten. The value is no longer entirely in the AI model, as that is gradually becoming a commodity. The value is in your data, your workflows, and your ability to orchestrate these cheap, powerful brains into products that solve real problems.

This brings us to a powerful historical parallel. The giants of our era, like the US rail tycoons of the 1800s, are spending astonishing capital to build the core AI infrastructure. They have laid the rail tracks for AI. But as investors in that era learned, the companies that laid the tracks were often the ones that went bankrupt. The great, lasting fortunes were not made by the rail companies, but by the thousands of businesses, the farms, factories, and new cities that sprang up alongside those tracks and railway stations. It's no longer only about who owns the rail tracks, but who can best ride those tracks to success.

CHAPTER 5

The AI Gold Rush: Opportunity or Illusion?

ChatGPT triggered the largest surge in history, driven by a single technology and the industries surrounding it. The most valuation, the fastest valuation to $100 billion, the largest data centers, the largest debt financing deals for hardware—the records that broke have been innumerable.

The verbal wars that are being fought over the future of AI and humanity have been intense, and some of the world's richest and most powerful have not hesitated to take sides. Some of the rhetoric even borders on fanaticism.

The DeepSeek moment, by itself, teaches us that bigger is not always better—that hype can go too far. It also teaches us that we may be arriving at a point in history where technological advantage may not always lie with a region or a country, that a host of nuclear non-proliferation treaties may not be needed (or enough!) to curb technology, that the democratization of power in the world may indeed be possible. That is a scary thought, but also an exciting one! As Satya Nadella once told us, you can always think about ten things that could go wrong, not work, or be bad. However, consider the question: "What if it works?!"[1]

---

[1] Personal conversation with the author, Monish.

© Harshad Oak, Monish Darda 2026
H. Oak and M. Darda, *The DeepSeek Moment*, https://doi.org/10.1007/979-8-8688-2598-9_5

There is something here in this age of AI that is different—that seems to defy what we have learned so far about technology and its effects. In this chapter, we delve into what is and could be different and what is and could be similar. We take those differences and similarities to draw out some ways to analyze this gold rush + illusion and hopefully learn to be better prepared in the process.

The "Mavericks" who believe AI will disrupt their industry and the "Custodians," the ones who believe that being the best in what they do is a necessary and sufficient condition for survival (you will get to know both better in a later chapter), should both benefit from the analysis presented in this chapter.

The Gold Rush

Nvidia's story has been incredible—a US$4 trillion (with a T!) increase in valuation, followed in the trillions with Microsoft and Google (Figure 5-1).

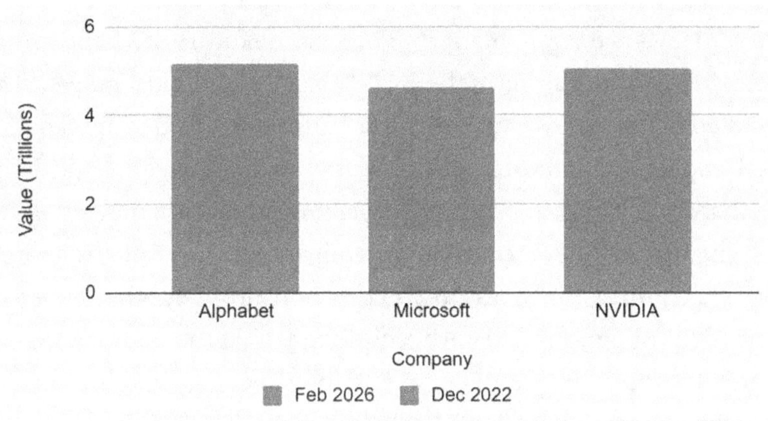

Figure 5-1. *Comparison of 2022 and 2026 market caps of Nvidia, Microsoft, and Alphabet*

OpenAI's valuation journey has been nothing but outstanding and unprecedented in many ways!

Talk about a gold rush! Compare them with Facebook or Amazon, and you will see what we mean. And it does not feel like any of the growth is slowing down.

There are some important observations:

- OpenAI's revenue was $3.7 billion in 2024—Amazon and Facebook were pulling in 10s, even 100s, of billions when they hit $ 500 billion in valuation.

- Hardware and compute innovation is going hand in hand with software innovation, and, of course, because hardware is much more expensive and capital-intensive, it is getting much more funding.

As shown in Table 5-1, the valuation of OpenAI expanded roughly 60-fold between 2021 and 2026, climbing from $14 billion to $852 billion. The company does not publicly disclose audited financial statements, yet reports suggest a 2024 revenue of approximately $3.7 billion and a December 2025 annualized run rate near $20 billion. The resulting valuation is roughly 40 times annualized revenue. A multiple of this magnitude is exceptionally high even for the software sector and implies that the market assumes a highly durable economic moat.

- Productivity software (Microsoft or Google compared with Salesforce or Workday) is part of our daily routine and has seen the most uptick.

- An interesting cycle has already started. The company with the most valuation in hardware (Nvidia) is funding the company with the most valuation in software (OpenAI) by renting its chips for the future.

Table 5-1. *OpenAI Valuation Milestones (2015 to 2026)*

Year	Event	Valuation ($B)
2015	Founding (nonprofit)	n/a
2019	Microsoft $1B Investment	not disclosed
2021	Employee Tender Offer	14
2023	Thrive / a16z Primary Round	29
2024	Thrive-led Primary Round	157
Mar 2025	SoftBank-led Primary Round	300
Oct 2025	Secondary Share Sale	500
Mar 2026	Primary Round	852

OpenAI, Microsoft, Google, and others are deliberately but surely moving into domains that were almost impossible to crack—OpenAI is moving into Microsoft's and Google's duopoly (or is it monopoly? ☺) of productivity and document management. Apart from fundamentally changing how people browse and buy things, OpenAI and companies like Anthropic and Perplexity are changing how we code, how we hyper-automate, and how we run consumer and enterprise workflows. Microsoft and Google are moving into drug discovery, healthcare, and almost every other industry, with substantially increased investments in their own models and data centers.

Apart from these companies that have already found gold—revenue and seemingly insane valuations—and the investors who have already cashed in on the boom, there is the rest of the world trying to find value in this post-DeepSeek scenario.

The Illusion

We have seen market exuberance defy logic before. Monish experienced this firsthand while living in the Bay Area during the dot-com boom. One story from that era stands out clearly. A friend asked for his feedback on a startup idea. She planned to create a website for organizing children's birthday parties. This was the early 2000s. The site would allow parents to create invitations, build a gift registry, and purchase party supplies. The business had no physical footprint. Monish warned her that execution would be incredibly difficult. He pointed out that the pricing model was weak and customer loyalty would be low. Acquiring users would require immense capital. He gently advised her against quitting her day job. One week later, she returned with a multimillion-dollar term sheet. Venture capitalists valued her company at over ten million dollars based solely on a slide deck. She did not even have a website then.

The point of this story is not whether the idea was good or bad—it showed that investors were so caught up in the Kool-Aid they were drinking and founders were so caught up in the opportunity that every idea seemed like a great one that would make them millions! Reality of course is very different. This seems to be what AI and AI companies are going through today—the illusion of revenue without value, or the illusion of value without the fit. We have also seen a more common pattern— expectations are so high because of the hype that use cases are warped to fit the technology, and when value is not delivered, the technology takes most of the blame.

Here is another example: most people who use ChatGPT for the first time get blown away by the seemingly superhuman responses. It feels like it can do anything and can think for other people. Then they hear and see the hype, and they are convinced it can do anything. So the obvious use case is that I have 200 people who review my financials—I will replace them with a trained LLM. It looks great on paper, and it delivers a great ROI—it also catches the attention of my CFO and my CTO, not to mention the CEO whose board is constantly asking: what are we doing with GenAI? Promises are made, plans are drawn, budgets are allocated, and a team is hastily assembled to show that this can be done in three weeks. The result? The project fails miserably, people lose trust in the team and technology, and sometimes morale goes down. Everybody starts talking about how this is an illusion, how 90% of AI projects do not move beyond the pilot phase, and that the technology will not be able to deliver anytime soon. We will address this more in our next chapter.

The difference in the case of AI has been that an almost magical technological advance comes along as soon as the trough of disillusionment is reached.

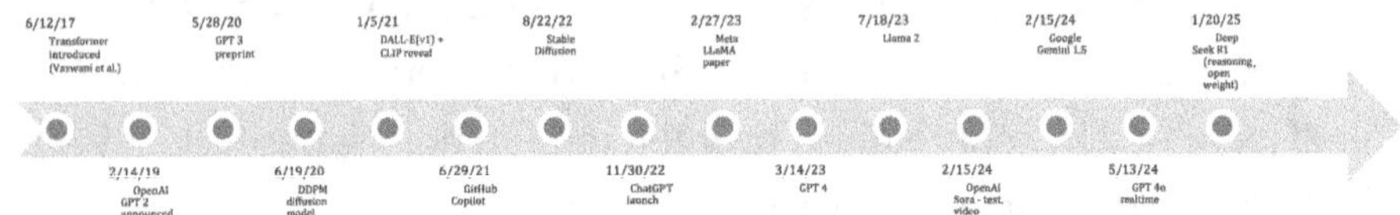

Figure 5-2. *Major milestones in modern AI*

Prompt-based image generation, prompt-based video generation, vibe coding, agentic workflows, the AI supercomputer on a desktop—it does not stop. That has raised the "anticipation premium (AP)" of these AI companies to unheard-of heights.

- **Anticipation premium is essentially "revenue realism × valuation humility."** If AP is tiny, investors are assuming *very* fat future revenues **and** tolerating high sales multiples today.

- **OpenAI vs. Nvidia:** Both show extreme "priced-for-tomorrow" footprints—one private, one public—but via different engines (subscription/API vs. silicon). That's a neat way to compare apples to GPUs. ☺

- **Alphabet's middle ground** suggests investors are optimistic but not hallucinating (we hope).

Execution Over Illusion

Market mania is a natural stage in every major technological shift like the one we are currently experiencing. However, beneath the financial noise, a true transformation is taking place. The DeepSeek moment has permanently lowered the barrier to entry. The anticipation premium will inevitably shrink. Boards and investors are increasingly demanding actual revenue and a clear return on investment.

The current illusion is that artificial intelligence will automatically fix broken business processes. However, technology is still a tool and not magic. The real opportunity lies in organizations going beyond experimentation and executing with purpose.

In the next chapter, we will explore how enterprises are making this transition. We will move from the theoretical gold rush to the practical reality of delivering value to customers.

CHAPTER 6

Enterprise AI in Action: From Contracts to Customers

Efficiency is the language of love for the enterprise; governance is the chaperone that lets the romance happen and AI makes it flower.

—An especially caffeinated CTO

The DeepSeek moment taught us that AI budgets don't have to break the bank and a talented, determined team of people on a (relatively) shoestring budget can work wonders. It gave us the ability to think bigger and, more fundamentally, about new applications and new ways of doing business. Though this empowered some enterprise leaders to think differently about their businesses, acceptance of their ideas and the execution of these ideas have remained a major challenge.

This has also led to some "expert" reports that applied AI is failing in the enterprise. Though there is some truth to that, it is also a function of how enterprises work in the real world. If you want to succeed in your

AI strategy, this understanding is critical. We took the liberty of offering some background and proposing a framework to increase your chances of success in the aftermath.

In this chapter, we delve into how we can adapt to this disruptive technology—grounding the reader in strategic truth. The core message is that awe gets you applause, but ROI gets you budgets—and delivering that ROI gets you success. But before we can, it is important to understand how enterprises work and which challenges are the most critical to overcome in the face of the disruption that AI brings.

The Elephant, the Roller Skates, and the CFO

Enterprises are unlike consumers: their appetite for AI is tempered by procurement, compliance, and return on investment (ROI). As enterprises face one of the biggest shifts in technology since electricity and the internet, it poses a challenge as big as teaching an elephant to roller-skate! From designing the right roller skates from scratch to overcoming fear and helping the elephant adapt, and from managing change so that the surroundings are protected from disaster, this enterprise elephant needs to learn to ride quickly, correctly, and safely.

The Nature of the Enterprise Beast

Enterprises work differently, and sometimes their behavior seems to defy simple logic. That is because they work differently, and unless there is a super-special leader at the helm (think Steve Jobs, Satya Nadella, or Sam Altman) or an incredible breakthrough, they tend to run by a loose, mostly difficult consensus. Here are some of the ways to break down this beast as you get ready to put enterprise AI into action:

- Multiple stakeholders

- Risk aversion

- Legacy systems

- Cultural resistance

- Global regulation

Let us look at them one by one.

Multiple Stakeholders

This is perhaps the biggest challenge for the enterprise. Getting people to agree to take a course of action, especially when the change is disruptive, is incredibly hard! Probably the hardest problem AI will need to solve eventually. Take, for example, Kodak. If asked individually, most people may have realized the threat of digital cameras. But as multiple stakeholders who had something to lose or change in chemicals, paper, hypo, film, cameras, distribution, and financial models, everything was being disrupted by digital, and consensus to do the things that would save the business must have been hard to come by.

When portable computers were a pipe dream, countless meetings must have concluded that portable computers were not really a threat, and no one would want a computer in their bag. Not because the people who participated in these meetings were not smart or did not realize the threat, but because it needed too much effort to drive consensus. So we come to the first lesson of the next generation of AI transformation:

"Do not run the AI transformation by consensus.

Get four or five of the brightest people with the strongest personalities together and give them leeway to make big decisions in controlled conditions."

Risk Aversion

A common belief even among those who know their science is that "Electric current only follows the path of least resistance"—no, it does not! Most current goes through the "least-resistance" path, but some current always goes through all the possible paths!

Enterprises are different. They usually choose the path of least resistance and rarely venture on the high-resistance paths. Risk aversion usually translates into downplaying the threat of new technology or a new manufacturing process because it has the potential to divert attention from existing business.

Microsoft's Steve Ballmer went on record ridiculing the "phone without a keyboard" removing the one threat to Apple's future dominance, and Kodak ignored their own invention of the digital camera to allow the Sonys and Canons to run away with an incredible market.

Even though it seems like it was a belief in their own strategy, the thinking was driven by the easiest path to take when resources were already committed. This is the second lesson of the next generation of AI transformation, and the second part is crucial:

> *"Build systems to take risks, and when they pay off, remove the resistance to embrace them."*

Legacy Systems

At Icertis, we encountered contract management systems at our customers that were decades old. Replacing them, even when the business realized the need, was one of the most difficult endeavors we had ever undertaken. This is one of the reasons why mainframe systems from the 1980s still run some of our most critical infrastructure. They are incredibly hard to replace because enterprise knowledge is buried in legacy, and when things are working, it is hard to stop.

So here is the next lesson:

"There is never a good time to replace systems.

Give them an end-of-life date (no more than a decade!) and then make it a corporate policy to replace them. And don't forget to budget that!"

Cultural Resistance

They say culture defines companies. At Icertis, even before we had a product, we articulated and defined our values and our mission that established our culture. But we never take it for granted, and we have had "re-founding" moments in our history where we have questioned our culture in a good way.

Building a culture that is not afraid of change, nay, embraces change, is the way the enterprise will be built in the new age. The culture of a company evolves—it has to get better, more inclusive, and more diverse as it matures. It needs to take into account the social and physical change people who make up the culture undergo.

Overcoming cultural resistance with a growth mindset is key to being a successful enterprise in the AI era. And this change can happen when you constrain resources. Look at how DeepSeek came to be—they had to ensure they challenged established beliefs because they did not have the resources to afford those beliefs, and so fundamental change in thinking and execution was the only route open. So here is the next lesson for modern enterprise:

"If culture gets used as a shield to resist the right change, be prepared to change the culture."

Global Regulation

Global regulation is only getting worse and more complicated. Global business depends on global regulation, and this adds a significant burden on the company. At the same time, it is necessary and has to be followed and, equally important, has to be backed by proof that it is followed.

One of the biggest fallouts of this regulatory environment is risk aversion—if you make a mistake, it can turn out to be extremely costly, sometimes even existential. But most regulation has rationale, and the way to do business is to ensure that people understand and, where they can, empathize with the rationale. And as Nassim Nicholas Taleb says, anti-fragility is where you take seemingly adverse conditions and turn them to your advantage—regulation can be a great example of anti-fragility. So here is the last lesson in the age of AI:

> *"As the regulatory environment (including AI regulation) becomes more complex, seek new ways of making it work for you."*

Dealing with Reality

As we did our research for this book (heavily supported by AI of course!), we started seeing some patterns in our failures and our successes. Here are some of the key patterns that stood out, and that helped us win, expand and delight customers, and hopefully keep us on track to be an even more consequential company.

Consensus Loves Certainty

Even with multiple stakeholders who are extremely risk-averse and bound by legacy systems and complex regulations, consensus can be driven when certainty of an outcome is promised and demonstrated. For a specific CLM implementation where AI would help draft engagement letters, an initial proposal from the project owner described that implementing this new system could potentially save the company a lot of money but it also meant a lot of change in how people work.

Though the promise of saving money was attractive, it led to a lot of seemingly never-ending discussion and debate—ranging from how to calculate the savings (that was probably more expensive than implementing the system itself 😊) to why it would not result in any savings at all.

One of the project sponsors made a simple tweak—she slashed next year's head count budget by 7% in her department because she believed the system would deliver those savings. This certainty, this clarity of thought, made stakeholders realize that this could be real. And she was not really taking a risk—the system was designed to provide a lot more savings!

But people love numbers and commitment—better still, they love action. And stakeholders, even when they are part of a large enterprise, are people! Stating clear outcomes and committing to those outcomes make the biggest disruptions seem easier.

Adoption Is 80% Org Chart

How many times have good systems been rolled out to die a dusty death? How many times have you seen people bypass good systems in favor of pushing a cart with square wheels uphill? Adoption is key, and adoption begins with the org chart! We have learned this the hard way—we have spent a lot of money, energy, and time building adoption into the platform, educating users on the benefits of adoption, and building consensus through internal marketing within our customers.

That is all necessary—it is not sufficient! Showing the CEO or the CPO using the system, reaching out to that hierarchy 1:1, and convincing them that it is important to show people that they believe in the change are critical. Aligning the org chart from top to bottom is super critical for adoption. Managing the org chart properly during an implementation can get enterprises to handle risk better and drive consensus quickly. And with AI, this is even more important—the disruption is not just at one layer, it is pervasive!

HitL Is Important, but HotL Is Critical

HitL, or human in the loop, has always been a critical part of transformation and change management. Ensuring that humans can take supplemental action when the system fails, or under circumstances where the system was not designed for, is critical to making things work. And with hallucinations, incomplete replies, and the very non-deterministic nature of AI, HitL becomes an important part of any AI transformation.

But what many people ignore is the importance of HotL—or human on the loop. The importance of "watching" what is happening and ensuring that things don't go offtrack—through dashboards, alerts and otherwise monitoring KPIs, high-value transactions, and trends—cannot be overstressed.

HotL allows transformative systems to succeed—picking up failure points, trends, and even human action that can be used to make the transformation succeed. Invest in HotL just like you would in HitL for your AI transformation. The fast-changing face of global commerce and regulation also requires HotL to be implemented right.

Riding the AI Wave

Given the nature of this elephant, conventional means, methodology, or philosophies may not be enough to ride the AI wave that has the potential of disrupting businesses in ways we never thought possible. To tackle this, we propose a flexible framework to tackle the first few years of this wave so that we don't get left so far behind that recovery is impossible. We also try and leave space in the framework for anti-fragility—being the disruptors yourselves as you work the AI wave.

Reorganize

You may have done reorgs before, but it is likely that this exercise may be unprecedented in the history of your company. This will be, by far, the most difficult step in your journey, but it will have to be done. Here is, very loosely, how we did it—you may have your own journey:

1. Have a strategy meeting on where we will land two to three years from now as a business.

2. Two clear camps emerge:

   a. One camp that predicts disaster if we don't embrace AI now and change the company

   b. The second camp that says that AI does not work yet, and will not disrupt the business, so let us focus on our core business while taking small side bets on AI

3. Separate the two camps—bound together, they will cause frustration and distrust.

4. Take the first camp (call them the Mavericks)—form a separate company if you must, but give them at least 10% of your top line as a budget, and get them to hire the right people to rethink the business from the ground up.

5. Take the second camp (call them the Custodians)— challenge them to use AI to compensate for the loss of the 10% from the core business's top line.

Here is our rationale for the process:

- People in the know realize the transformative power of AI—the rest of the world will almost always underestimate it. Satya Nadella—almost overnight—handed over the commercial reins of Microsoft to Judson Althoff making himself focus completely on Copilot and Microsoft's AI strategy. Not because he was bad at it, but because he realized that for Microsoft's future, focus on the AI bet is critical.

- The bet has to be healthy; weak bets will not allow you to survive a real disruption. So the division of your leadership team and 10% of your top line—use the numbers that make sense to you, considering that this could be the difference between survival and an inconsequential existence.

- Create scarcity in the rest of the company to help become more efficient and to make people adopt AI for efficiency.

Focus

People, especially the engineers, can quickly get carried away by the technology. The Mavericks have to be guided, and they must be squarely focused on one thing—business outcomes with AI. We propose three principles:

- Identify the top five use cases in the business. Five is a number that seems to work well, but you should choose your own. You don't want too many—focus is critical. You don't want too few; you will reduce your chances for success.

- Create a three-year plan for each of these use cases, starting with progressively increasing outcomes.

- Plan for the outcomes to be substantial; else, revisit the use cases. Remember you are playing with 10% of the top line; the outcomes must make the bet worth placing!

Make the technology work for the use cases, with clearly defined business outcomes. That way, every shiny toy out there will not be a creator of new use cases at the expense of the ones that have already shown some progress. Focus helps commit long-term and makes teams more resilient to failure—and fail they will—before they can taste success. So it is important to focus and be resilient to failure in the short-term.

Educate

Everybody seems to agree that change management is one of the most difficult parts of running an enterprise. Given that AI may possibly be the biggest change driver in the history of your company, education is key. At Icertis, we pushed an AI education agenda very aggressively. We conducted brown bag sessions, encouraged Icertis to publish stories of their AI adoption internally and externally, created awards for people who built AI use cases, and invested in GitHub Copilot and Microsoft Copilot licenses even at the time when the return on investment was unclear. We did that to get people used to the idea of what AI could do to their lives and to get them to use it even when it worked imperfectly. That is key— the early adopters and companies who end up staying ahead of the game are the ones who show patience when the dust is settling and the hype is dying down.

Education did not end with Icertis. Every customer connect event has discussed the pros and cons of AI, what works and what does not, and just the importance of using AI even if it is imperfect. We steered conversations toward the idea that reducing costs may be the immediate step, but transforming the business must be the eventual goal of any AI strategy. And we educated analysts (and learned from them) on how contracts would play a critical part in any AI strategy—after all, everyone agrees that guardrails are absolutely necessary for AI, and where are the guardrails and rules of business enshrined? They are in your contracts!

Education also did not end with others. There was and is a continuous desire and a systematic plan to educate ourselves. Carve out 15% of time to spend on reading, learning, and talking to customers, CxOs, technologists, investors, financial experts, and Gen Z to get closer to the shape of the technology and what patterns were emerging. It has been an incredible journey of learning and revelation. This book has been an outlet for that incredible journey.

Governance and Monitoring

The Mavericks will not just want the elephant to skate—they will want it to fly! The Custodians will want to keep the elephant healthy so it can keep walking! Both may be doing the right things for the company, but they are optimizing for themselves and what they believe in. So it is imperative that your governance is of the highest quality. KPIs have to be set, monitoring has to be tight, and the outcome of non-conformance has to be ruthless.

I remember reading about how Satya ran (and probably still does) these weekly meetings in the early days of Microsoft Copilot from marketing to support and everything in between when revenue was not even in the 10s of millions because he had realized then that this was the engine that the company would transform around.

High governance and monitoring help with the Mavericks as well as the Custodians (see Table 6-1). Making sure both sides are tightly governed needs a skilled project executive—in most companies, this is a critical hire at this stage because this talent is hard to find!

Table 6-1. *Governance Matrix for Mavericks and Custodians*

	Low Governance	**High Governance**
Mavericks	Chaos Zone—rogue pilots, wasted spend, compliance risks	Innovation Zone—balanced bets, coordinated scale, safe disruption
Custodians	Dormant Zone—stagnation, missed upside, talent attrition	Discipline Zone—steady ROI, slower pace, technology readiness

Tight governance also means that dates and milestones are considered sacrosanct. Sometimes, killing a pilot because it could not meet a milestone sets a good example and drives discipline. Ruthlessness helps—it is a sign of efficiency.

Pitfalls and Counters

This is probably the most intuitive and probably a safe-to-skip section if your company is already run well. Table 6-2 is included for completeness.

Table 6-2. *Pitfalls*

Pitfall	**How It Shows Up**	**Countermove**
Fanaticism (Chaos Zone)	"AI everywhere by Friday!", duplicated spend, inconsistent standards.	Institute a use case funnel and central asset reuse.
Orphaned pilots	Projects launch with fanfare, never scale.	Require integration and ownership plans before kickoff.
Governance theater	Lots of paperwork, little enablement.	Measure time-to-yes, publish playbooks, lean policies.
Change fatigue	Clunky tools, thin training, low adoption.	Invest in UX, champion networks, reward adoption.
Regulatory misses	Privacy lapses, explainability gaps, audit pain.	Engage counsel early, map controls, ensure traceability.

The Path Forward

Teaching the enterprise elephant to roller-skate may seem silly, but it is a great metaphor for the magnitude of the problem you face. The middle road between the Mavericks and the Custodians is a difficult road to tread, and you will need new skills, new ways of thinking, and the discipline to question what has worked for decades, while being pragmatic about what you can expect. We hope that this framework helps you think about the problem and evolve your own. We would love to hear and learn about and from your unique challenges.

CHAPTER 7

Building AI You Can Trust: Ethics, Accountability, and Real-World Choices

It is change, continuing change, inevitable change, that is the dominant factor in society today ... No sensible decision can be made any longer without taking into account not only the world as it is, but the world as it will be.

—Isaac Asimov, in his essay "My Own View"

As we saw in an earlier chapter, AI has the potential to be the future of humankind. The DeepSeek moment has shown that the pace of change has been and will be unlike anything the world has ever seen. In this context, taking into account the world as it will be is critical. In this chapter, we discuss the challenges of ethics and accountability and the pragmatic real-world choices that must be made today and into the future.

© Harshad Oak, Monish Darda 2026
H. Oak and M. Darda, *The DeepSeek Moment*, https://doi.org/10.1007/979-8-8688-2598-9_7

The Ethics of AI

Humanity learns through experience. We have developed our ethical frameworks over centuries of trial and error. Yet, history shows that greed and power often cause us to abandon this understanding. The real-world choices we make today will determine whether AI fuels that notoriety or pulls us out of the depths that we plunge ourselves into occasionally.

The Application to Robotics

If AI is the brain, a robot combines that brain with action in the physical world. We go back to Asimov's seminal laws of robotics that become so relevant to the evolution of this technology:

- **The First Law**: "A robot may not injure a human being or, through inaction, allow a human being to come to harm."

- **The Second Law**: "A robot must obey the orders given to it by human beings except where such orders would conflict with the first law."

- **The Third Law**: "A robot must protect its own existence as long as such protection does not conflict with the first or second law."

These laws governed the robotic universe of Isaac Asimov of a future in which robots and humans worked together for the betterment of humanity. As Asimov explored this world, his robots faced bigger and bigger challenges in implementing these laws (as you can see, these laws are simplistic in the real world)—here are some simple ethical dilemmas that created a paradox as examples:

- If the robot could only prevent a human from killing several other humans by causing injury to the human, would the robot fail to prevent the killing because it could not cause injury according to the first law?

- If a robot obeyed an order without knowing that obeying that order would end up injuring other human beings, how would it ever make the decision to obey any order once it encountered the first order that resulted in injury to a human being?

You get the idea—these laws have the right intention, but they are sometimes impossible to implement in the real world. Asimov's robots found a way—they *altered their programming themselves* to incorporate a Zeroth law:

> **The Zeroth Law**: "A robot may not harm humanity or, by inaction, allow humanity to come to harm."

By ensuring that the good of the many came before the good of the few, it attempted to bring context to decisions that were always going to be hard within any framework.

The Ethics of Autonomous Cars

The closest manifestation of the AI-driven future today is the autonomous car—if you have ever experienced Waymo or Zoox, you will instantly relate to the dilemmas that scientists, implementers, governments, and laws had to solve to make today's autonomous cars practical.

Assuming the three laws of robotics loosely hold for autonomous cars (do not harm a human being, obey the human being within reason, and protect itself from harm within reason), we quickly see how the ethics need to evolve:

Should the car be over-cautious, making San Francisco's streets impossible to navigate at peak times? Or should it be practical, allowing some risk? Asimov's first law collapses in continuous control systems. When you experience the autonomous car, you realize that the driving patterns are very similar to an average driver in San Francisco, within the limitations of the law. The risk and rewards have been balanced to tune the outcome.

For the car to obey humans, it has to understand laws, speed limits, lane controls, and traffic lights and be as aware of its environment as possible. Autonomous cars take the simplest route—they constrain human commands to a handful (play music, adjust air conditioning, pull up, call emergency services, etc.) so that risk does not increase. By ensuring that the car has built-in knowledge of traffic rules and is constrained to follow them, the risk is reduced, if not eliminated. But because the autonomous car is still substantially safer than an average human driver, that risk is then accepted by the law that allows autonomous vehicles to co-exist with humans on the road.

The second law thus becomes a negotiation between human, machine, and the law.

The third law is slightly easier—within the constraints of the first two, the car will try and protect itself.

With this background, for ethics to work with AI, some of the following problems will need to be solved:

- **Which Use Cases Are Off-Limits for AI and Will Always Require a Human Being?**

  - When a human's life is in balance: Murder trials, for example, may only be judged by humans, not AI. Or a drug that can be fatal but has the potential to cure a potentially fatal disease can only be administered by a human.

  - Figuring out where a human in the loop is essential and where it is optional will be critical.

- **What Data Is Used to Train?**

  - AI depends on what has happened in the past to
    provide answers to problems that may arise in the
    future. If this past has bias, that bias will reflect in the
    inference that may then be deemed unethical
    by humans.

  - Ensuring that data is as clean and unbiased as
    possible will be a significant problem to solve.

- **Moral Standardization Across Cultures**

  - This is interesting even in the context of today's
    regulations like the General Data Protection
    Regulation (GDPR). If a set of countries provides
    strict regulation to protect personal data, and
    another set does not care, do the countries that do
    not care have an unfair advantage if they compete in
    the same markets?

  - Moral standardization across cultures is difficult to
    imagine, but it will be a necessary condition for the
    adoption of AI in every aspect of human life across
    the world.

- **Explainability and Responsibility**

  - As the technology evolves, ensuring that inference is
    explainable (what were the steps involved in reach-
    ing a conclusion, what data was used to make deci-
    sions, what references and legal or social
    benchmarks were used, etc.) will need to be solved in
    a much better way.

- **Accountability**

    - As AI robotics takes even more actions in the real
      world and impacts human life significantly, the
      question of accountability will need to be resolved.
      Especially for scenarios where a human is not in the
      loop for action through AI inference, accountability
      will be critical. Some parts of this problem have been
      addressed with the accountability of actions by
      autonomous vehicles, but there is a long way to go.

The Evolution of AI Regulation

Some of these challenges have already been taken up by AI regulation that
has been or is being drafted across the world. Let us look at the European
Union (EU)—the EU has led the way in data privacy and personal data
protection, and is also the first region to adopt a law that governs the use
of AI. The central idea is "the more dangerous an AI system could be, the
stricter the rules." It sorts AI into four buckets:

- **Unacceptable-Risk AI (Use of AI Is Banned)**

    - Social scoring of citizens

    - Manipulative AI exploiting vulnerabilities

    - Legal judgment (e.g., in a murder trial)

- **High-Risk AI (Allowed but Heavily Regulated)**

    - AI in hiring, education, healthcare, credit scoring

    - AI in infrastructure

    - AI in border control systems

- **Limited-Risk AI (Light Rules)**

    - Chatbots, recommendation systems

- **Minimal-Risk AI (No Extra Rules)**

    - Games, spam filters, photo editors, etc.

It is the world's first major AI law, which took six years of political and social debate, ethics guidelines, and the rapid progress of technology. Much like the laws of robotics, it is not perfect and certainly not easy to implement broadly, but it is a start.

The world has followed, but the complications are immense. For example, the United States has a set of executive orders and some sectoral laws that cover autonomous vehicles (DoT), healthcare (FDA), consumer protection (FTC), and border security and immigration (DHS). The United Kingdom has a pro-innovation framework, with a "principles based, regulator led" model and the world's first AI Safety Institute. China has taken the path of algorithm and AI regulation—platforms like Tencent, Alibaba, and Baidu have to register their algorithms with the government according to the Algorithm Recommendation Regulation of 2022, mandatory watermarking and real-name verification for deep synthesis/deep fakes, and generative AI measures of 2023.

For reference, Table 7-1 provides a loose (based on our interpretation) of the AI regulations across the world.

Table 7-1. *AI Regulations*

Region/ Country	Type of Regulation	Key Features	Status (February 2026)	Philosophy/ Approach
European Union (EU)	EU AI Act (comprehensive law)	Risk-based tiers, bans, strict rules for high-risk AI, transparency for chatbots and deepfakes, rules for foundation models.	In force (phased implementation 2024–2027)	Rights-first, precautionary, safety-focused
United States	Executive order + sectoral laws	AI safety tests, watermarking, red-teaming; regulation via FDA, FTC, DOT, etc.; state-level laws emerging. Export controls on advanced AI chips.	Active but no single AI law	Innovation-first, regulate by sector
United Kingdom	Regulator-led principles	No single law; guidance via ICO, CMA, FCA; AI Safety Institute; testing for advanced models. Online Safety Act obligations apply to AI chat platforms.	Framework in place	Innovation-friendly, flexible

(*continued*)

Table 7-1. (*continued*)

Region/ Country	Type of Regulation	Key Features	Status (February 2026)	Philosophy/ Approach
China	Strong operational rules	Algorithm registry, deepfake watermarking, real-name verification, security reviews, generative AI controls.	Generative AI and deep synthesis regulations in force and enforced	National security + social stability
India	IT Amendment Rules 2026 and DPDP Act	DPDP Act (data governance); IT Amendment Rules (platform liability, deepfake controls, due diligence); government advisories on generative AI; responsible AI initiatives.	Active (DPDP Act operational; IT Amendment Rules effective February 2026)	Enable innovation, protect users

(continued)

Table 7-1. (*continued*)

Region/ Country	Type of Regulation	Key Features	Status (February 2026)	Philosophy/ Approach
Singapore	Governance framework (non-binding)	Model AI Governance Framework, testing toolkit, sandboxes, explainability and transparency guidance.	Guidelines active	Business-friendly, voluntary compliance
Japan	Guidelines (non-binding)	Ethical AI principles, business-use guidelines, focus on trustworthy AI and social harmony.	No dedicated AI law	Minimal regulation, self-governed
UAE	National strategy + sector laws	Minister of AI, guidelines for responsible AI, AV regulations, gov-led AI scaling.	Active	Become global AI hub, pro-innovation
Canada	Proposed AIDA (AI and Data Act)	Regulates high-impact systems, risk mitigation, transparency, heavy penalties for misuse.	Proposed but not enacted. AI Safety Institute launched.	Align with EU, protect consumers

(*continued*)

Table 7-1. *(continued)*

Region/ Country	Type of Regulation	Key Features	Status (February 2026)	Philosophy/ Approach
South Korea	AI Basic Act	Mandatory labeling for generative AI and deepfakes, heightened transparency/safety requirements for high-impact AI fields, AI sandboxes.	In force (January 2026)	Balance innovation + safety

Ethics and AI Regulation

Ethics and AI regulation go hand in hand. There are major problems to
be solved, but it is encouraging that in many areas including autonomous
vehicles, we are seeing a rapid evolution of regulation with the ethics of
use and accountability being addressed in pragmatic ways. Regulation
will always be outpaced by innovation, and ethics will sometimes seem
secondary in the rush to lead, but we are very optimistic that the human
race will figure this out for the betterment of all.

Real-World Choices

*If knowledge can create problems, it is not through ignorance
that we can solve them.*

—*Isaac Asimov's Book of Science and Nature Quotations* (1988)

With so much happening and the burden of regulation looming over businesses, it is fascinating to see how commercialization is taking place. In this section, we explore how you, as a business owner, technologist, executive, or policy maker, handle the ever-evolving ethical and responsible AI landscape while still preserving your sanity and the competitive edge of your business.

We propose a framework that allows real-world implementations to become more robust while remaining agile and helps organizations prepare for the future. In proposing the framework, we seek a balance of pragmatism, responsibility, and accountability.

1. Bring organizational values and policies to bear on AI.

   a. Ground everything you do with AI in your organizational values.

      i. Let everyone in the organization know (several times, emphasizing with new hires using special training programs) that your values are important in everything you do, especially AI.

      ii. Ensure your values cover fairness (diversity in your data, teams, points of view, among other things) and respect (for your customers' data, for copyrights and creative art rights, among other things).

   b. Ensure you have a responsible AI policy that everyone knows about.

      i. Tie it back to organizational values and ensure you engender respect within the organization for the policy.

      ii. Clearly articulate what you will not do, no matter the commercial consequences.

   c. Review the implementation of your values and policies
      regularly.

   d. Review your values and policies regularly to ensure that the
      world has not left them behind (this will happen more often
      than you think!).

2. Bridge the hallucination–accountability gap as an
   integral part of your AI solutions.

   a. "Trust but verify" does not really scale when AI works 24×7.
      Ensure that inferences are always verified and the verification
      itself is attached to the inference as an audit trail.

   b. Always label AI output, decisions, and actions: this seems
      obvious, but is ignored many times.

      i. Make this part of your user experience tooling, automation,
         and test cases.

3. Beware of bias.

   a. Bias is an integral part of historical data in most organizations
      and, in some cases, unavoidable. But do not let this argument
      deter your organization from making an effort.

   b. Ensure that there is pragmatic testing for bias and top-level
      executive clearances that the best effort has been made by the
      organization.

   c. This not only protects you and your customers, it helps build
      better systems.

4. Control shadow AI carefully.

   a. You do not want to muzzle innovation, but you do not want to
      take your most sensitive data and throw it into the wind.

    b. Like data and app security, this starts with people, processes, and tools. Invest in all three and ensure that shadow AI is controlled in the organization and is restricted from using sensitive data till systems and safeguards are proven.

5. Data governance.

    a. This forms the foundation—knowing where your data is, its semantics, importance and sensitivity, how it is used, and where it is used is critical.

    b. Ensure you invest in data governance from the ground up, and make this part of your AI strategy.

6. Geopolitical compliance to regulation, policy, and law.

    a. Make sure there are processes in place to regularly review changes in compliance requirements in every geography you operate in.

    b. Loosely coupled systems with architecture that can turn features on or off as configuration are critical. Design your AI use with the right set of controls so that you don't wind up in all-or-nothing situations.

7. Human in the loop (HitL) and human on the loop (HotL).

    a. Build human intervention (human in the loop) and human monitoring (human on the loop) into AI systems and their outcomes.

    b. Recognize that products and services now need to be designed differently—just like autonomous cars, you will most probably end up owning (and/or taking responsibility for) the entire stack. Be prepared for the scale and responsibility and educate customers about the costs.

8. Make measurement a long-term, continuous feedback goal.

   a. This includes measurement of

      i. Hallucination percentage

      ii. Fairness and bias

      iii. Interpretability and explainability

      iv. Safety thresholds, including content safety

      v. Audit efficacy

      vi. Adherence to regulation, policy, and law

9. Like a good feedback loop, do not forget to leverage AI to implement ethics and accountability in your systems.

   a. Sometimes (and we know it is a bad metaphor), you have to set a thief to catch a thief. Ensuring that you get AI on your side will make your systems more robust.

This, as you may have already realized, is not a comprehensive list. But it is a complicated and sometimes overwhelming list. Ethics and accountability are not easy—for enduring organizations, they are built into the organization's fabric. If you do not embrace the responsibility, you can be rest assured that it will one day be forced upon you—and then you will not be prepared. So, as Asimov suggested many decades ago, ensuring that you take into account the world that will be—today—will increase your chances of success in the future.

Trust Is Your Moat

The DeepSeek moment proved that raw intelligence is increasingly a commodity. Soon, your artificial intelligence may not be smarter than the system your competitor uses. The core technology may equalize across the market.

When intelligence is equal, trust becomes the final currency.

Customers will choose the systems that guarantee safety, privacy, and fairness. Trust is your ultimate competitive advantage. The nine steps we outlined will help you build that trust within your own walls.

However, we must look beyond your corporate walls. The democratization of artificial intelligence has alerted governments worldwide. Nations realize that whoever defines the ethics of artificial intelligence will define the global economy. They are now competing to write the rules for this powerful new commodity.

We are moving rapidly from corporate accountability to national sovereignty. In the next chapter, we explore how this geopolitical shift is redrawing the map of global technology dominance.

CHAPTER 8

Redistribution of Intelligence Through Innovation

AI has often been framed as a story of technological progress defined by faster chips, larger models, and larger datasets. However, when a technology becomes powerful enough, it does more than rewrite engineering assumptions. It rearranges the world around it, much like gunpowder, electricity, nuclear power, or the internet.

What began as a race to build better models has turned into a quiet force shaping economies, national strategies, and the flow of innovation itself.

By the time the DeepSeek moment unfolded in early 2025, most leaders believed that the future of AI would follow familiar gravitational lines. A handful of regions would generate breakthroughs, while the rest of the world would consume them. This unipolar view assumed that high intelligence required high capital. It relied on the premise that the path to general intelligence was paved with tens of thousands of GPUs, multibillion-dollar data centers, and energy contracts that rivaled the consumption of small nations.

The efficiency breakthroughs described in Chapter 2 shattered this consensus. Suddenly, capability did not sit neatly atop capital. It expanded outward, carried by cost reduction, smarter engineering, and the appeal of independence.

© Harshad Oak, Monish Darda 2026
H. Oak and M. Darda, *The DeepSeek Moment*, https://doi.org/10.1007/979-8-8688-2598-9_8

We saw this redistribution take shape at the AI Impact Summit in India in February 2026. The event focused on the theme of welfare for all and demanded equitable access to computational and research resources. The Global South is no longer just a consumer of technology. It is actively demanding a voice in how global innovation is governed and distributed.

This chapter explores what that redistribution means. We look not merely at the language of geopolitics, but at the practical terms that matter to leaders: procurement, data sovereignty, and operational resilience.

Hardware Moat vs. Software Flood

For years, the prevailing theory was that the "moat" in AI would be built of silicon and concrete. The assumption was that as models grew larger, the capital required to train them would become so astronomical that only a few entities could compete. We now see two diverging philosophies challenging this assumption.

To understand the current rush for infrastructure, it is helpful to look at the Manhattan Project, the secret World War II undertaking that produced the first nuclear weapons. It remains the template for rapid industrial-scale mobilization to harness a new technology. Operating at the height of World War II, from 1942 to 1945, the project employed a peak workforce of 130,000 across 37 facilities and incurred an estimated expenditure of $2 billion (~$36 billion in 2026 dollars). The project was defined by the belief that a transformative breakthrough required the centralized coordination of the nation's best minds, unlimited capital, and scarce resources at a scale no private entity could manage alone.

Today, as nations and corporations view AI as a transformative economic asset, we see a similar logic emerging. This perspective treats computing like a strategic national resource that must be refined at an industrial scale to ensure leadership. The focus is on centralization and raw capacity. The logic here is that the barrier to entry is the price tag. By raising the cost of participation to hundreds of billions of dollars, the goal is to build a capability moat that cannot be crossed simply by being clever.

The Bet on Infrastructure

This philosophy is evident in initiatives like Stargate, the commercial consortium announced by President Trump in early 2025 and involving OpenAI, SoftBank, and Oracle. With potential investments reaching $500 billion, Stargate represents the ultimate bet on the "Scaling Law," or the belief that the path to Artificial General Intelligence (AGI) requires progressively larger models trained on progressively massive clusters.

The Bet on Open Replicable Software

DeepSeek represents the alternative argument. While the creation of a frontier model requires a massive physical footprint (the uranium), the finished product (the model weights) is purely information. It is a digital file that can be copied, leaked, and distributed instantly.

DeepSeek's efficiency breakthrough undermines the premise of containment. By prioritizing architectural elegance, they proved that intelligence does not require a massive physical supply chain to replicate.

Access Zones

While software travels at the speed of light, hardware travels on ships and trucks. The efficiency breakthroughs proved that code is fluid, but the physical infrastructure of intelligence remains stubbornly solid. For the global enterprise, this creates a new domain of risk.

Export controls and trade alliances have effectively sorted the world into three zones of access:

- **Zone 1—the Trusted Core**: Includes the United States and key allies. These nations enjoy access to the most advanced hardware. Supply chains are secure, but costs are high.

- **Zone 2—the Conditional Middle**: Includes regions like the Middle East, India, and others. Access to data center–class hardware is possible but subject to strict licensing and monitoring.

- **Zone 3—the Restricted Zone**: Includes China, Russia, and sanctioned entities. Access to advanced AI inputs is subject to an initial assumption that it will be denied. Enterprises here must rely on the domestic stack or optimized software to compete.

For the modern CIO, the strategic implication is blunt. You cannot assume that the hardware you order today will be legal to ship tomorrow. If your primary model provider relies on a supply chain that crosses a restricted border or if your backup data center is in a conditional zone with revoked licenses, you do not have an IT problem. You have a geopolitical continuity risk.

AI Hubs

When innovation becomes affordable, it stops belonging to any one place. The "DeepSeek effect has triggered a global race for Sovereign AI. Nations are no longer content to import intelligence; they want to generate it, aligned with their own laws, culture, and economic needs.

China: Creating an Artificial Demand

China faced a hard ceiling on hardware access due to export controls. In response, the region pivoted to a systemic focus on software architecture. DeepSeek is the initial proof point of this adaptation, having trained its V3 model on restricted Nvidia H800s.

However, the transition to full silicon sovereignty has introduced a period of intense friction. Reports indicate that DeepSeek's successor model faced delays when shifting training to domestic Huawei chips, eventually forcing a fallback to Nvidia clusters for some workloads.[1]

Crucially, Beijing views this not as a failure, but as a kind of "sovereignty tax." The government's strategy is to force an adoption cycle. By mandating that state-backed data centers and champions like DeepSeek use domestic silicon despite its current immaturity, they are creating a guaranteed market for companies like Huawei. This artificial demand provides the massive data feedback loops required to mature the software stack.

India: AI As Digital Public Infrastructure

While the United States and China compete to build the highest "walled gardens," India is building a public park.

India views AI not as a product to be sold, but as Digital Public Infrastructure (DPI), a utility akin to water, electricity, or India's Unified Payments Interface (UPI) that handled over 22 billion transactions in just the month of April 2026. UPI transaction counts exceed those of the global Visa network.

Having successfully engaged over a billion citizens with the "India Stack" (Aadhaar for identity, UPI for payments), New Delhi is now architecting an "AI Stack" to democratize intelligence at a population scale. The New "AI Stack" unbundles the value chain and intends to make intelligence accessible to the smallest startup or a farmer in a remote village.

---

[1] https://www.ft.com/content/19013793-4e8d-444b-b57f-5435352a8f3b

The Gulf: From Oil to Compute

The Middle East is executing a pivot from hydrocarbons to computing. Nations like the UAE and Saudi Arabia are leveraging their greatest assets, cheap energy and sovereign capital, to build the physical backbone of the global AI economy. Initiatives like Saudi Arabia's Humain, the UAE's Falcon model, and the global AI investments by funds based in the Middle East represent a bid to become the neutral ground of AI. They are also betting that while code may be efficient, the energy to run it will continue to be a major constraint.

Japan: Keeping It Domestic

Japan has funded domestic models because it views importing American AI as importing American norms. The government warns that models trained on English data tend to provide responses based on Western values with biases on hierarchy, history, and social harmony that may be incompatible with Japanese society.

Beyond developing indigenous AI models, the government is deploying state-backed evaluation systems to assess AI credibility. These systems test models against specific criteria, including whether their content aligns with Japanese culture and avoids foreign bias on sensitive topics.

South Korea: Thanks for the Memories

South Korea leverages its dominance in memory chips to secure a place at the table. Companies like Samsung are building full-stack AI ecosystems. The recent government-backed deal to secure 260,000 of Nvidia's Blackwell-class AI chips[2] highlights its intent to maintain competitiveness.

---

[2] https://www.reuters.com/business/media-telecom/nvidia-supply-more-than-260000-blackwell-ai-chips-south-korea-2025-10-31/

Europe: Riding the Trust Wave

Europe largely missed the cloud wave, but it is determined to win what one might call the trust wave. Through frameworks like the EU AI Act, Brussels is betting that regulation is a feature and not a bug. However, Europe is also waking up to the infrastructure deficit. New initiatives like the one to build Europe's largest AI campus in France are funneling billions into sovereign compute, realizing that you cannot regulate what you do not own. Companies like Mistral are spearheading Europe's build of frontier models that respect privacy and copyright by design.

Open Source As a Diplomatic Lever

In this fragmented world, open source AI has transformed from a developer tool into a diplomatic lever. For many nations like Canada, Israel, France, and the UAE, open weights are a strategic counter-balance.

By releasing models like the UAE's Falcon or France's Mistral, these nations achieve two goals. First, they break the US and Chinese tech monopolies. Second, they export their standards. When a developer builds on top of Mistral, they are implicitly opting into a European regulatory framework rather than an American one.

For the enterprise, this creates a new menu of options. You are no longer forced to choose between a US and a Chinese model. You can choose a "neutral" open model, host it in a jurisdiction of your choice, and fine-tune it on your own data. This is the "Clone and Localize" strategy: achieving 90% of frontier performance for a fraction of the cost, while maintaining data sovereignty.

AI operational resilience is the ability to ensure that critical intelligence cannot be disabled by a foreign sanction, a change in a vendor's terms, or a geopolitical blockade. The risks are not theoretical. In a stark example of geopolitical spillover, Microsoft Azure suspended services to Nayara

Energy,[3] an Indian oil refiner, citing EU sanctions due to the company's Russian backing. If your intelligence layer resides on a foreign cloud, your operations are subject to foreign policy.

Managing Enterprise Risk

Navigating this fragmented map requires more than just awareness. It requires a shift in how organizations procure, deploy, and govern intelligence.

1. **Inventory Your Geopolitical Exposure:** Ensure your critical intelligence supply chain does not pass through a single choke point.

2. **Pilot "Edge First":** Test a distilled model on consumer hardware today. The future is hybrid, and customers will expect intelligence to work using fast, lightweight, and on-device models.

3. **Audit for "Model Residency":** Data residency is no longer enough. You need to know where the model's weights reside and who controls the "off" switch. Ensure you have a sovereign fail-safe for mission-critical apps.

4. **Build a Switching Layer:** Agility is the only defense against fragmentation. Ensure your applications can swap LLM backends quickly. If one provider changes terms or faces a regulatory ban, your business must be able to reroute intelligence as easily as it reroutes internet traffic.

---

[3] https://www.reuters.com/world/india/russia-backed-nayara-taps-indian-it-firm-after-microsoft-suspends-service-2025-07-29/

5. **Secure the Model Supply Chain:** Operational security must now include the place of origin of the AI model. Attacks may manifest as poisoned training data or compromised weights. For regulated sectors, you must verify the integrity of the weights you download.

The Agentic Fracture

The era of the "AI Empire," in which a single company or a single nation ruled them all, is over. We are entering the era of the AI Federation. It is a messier world. It is fragmented, regulated, and complex. However, it is also more resilient. The leaders who win in this new world will not be the ones who simply pick the "best" model. They will be the ones who master the routes between them.

We must also prepare for the next challenge. In the next chapter, we will see how this geopolitical fragmentation accelerates as AI evolves from static models into autonomous agents. If we think managing a fragmented map of models is difficult, wait until we must manage a global economy of independent AI agents.

Consider an AI agent from the United States negotiating a supply contract with an AI agent from the EU. The US agent is optimized for speed and profit; the EU agent is constrained by privacy laws and carbon caps. They do not just speak different languages; they operate under different realities. The "DeepSeek moment" was the beginning; the "Agentic Moment" is where the complexity begins.

CHAPTER 9

The Agentic Shift and the Emergence of Service-as-a-Software

History has a habit of repeating itself. This often happens when the price of a core resource collapses. Consider the story of steel. Before 1856, steel was incredibly expensive and used sparingly, reserved for luxury items like swords and cutlery. Then Henry Bessemer invented a process to blow air through molten iron, and the cost of producing steel collapsed by over 80%.[1] The result was profound. We did not just make cheaper cutlery. We built skyscrapers and suspension bridges. Structures that were physically impossible suddenly became possible and viable.

Before DeepSeek, AI reasoning was our steel. It was precious and rationed, used only for short summaries and simple chatbots. DeepSeek collapsed the cost of reasoning.

In January 2025, DeepSeek shocked the global markets. Just over a year later, in February 2026, Claude Cowork from Anthropic triggered a selloff across many software and IT consulting companies on global stock markets. Claude Cowork showed how its agents could autonomously navigate file systems and execute complex workflows, bypassing

---

[1] https://www.thechemicalengineer.com/features/cewctw-henry-bessemer-man-of-steel/

© Harshad Oak, Monish Darda 2026
H. Oak and M. Darda, *The DeepSeek Moment*, https://doi.org/10.1007/979-8-8688-2598-9_9

traditional interfaces and requiring no manual guidance or supervision. The disruption was clearly spreading and moving up the technology stack. The Agentic Era is not a vision of the future. The agentic disruption is already at the application layer.

Why Efficiency Fuels Agency

To understand why this shift is happening now, we have to look at the math. But first, we must define what we mean by an AI agent. While a chatbot answers a question or writes a poem, an AI agent is a worker. It pursues a complex goal autonomously. Agents are different; they don't just talk, but they plan, act, reflect, and iterate until the job is done.

This distinction is critical because agents are computationally expensive. Imagine a complex task that requires the AI to check a database, write code, test the code, fail, fix the code, iterate, and report back when done. That might take 50 internal steps.

At legacy prices, like the early GPT era, a single task like that could cost up to a dollar. That does not sound like much, but if you run it at enterprise scale millions of times a day, the total bill becomes prohibitive. At post-DeepSeek prices, using that same loop now costs pennies.

DeepSeek changed the fundamental math. By driving the cost of reasoning down to near zero, it unlocked the agentic loop. When reasoning is cheap, you do not have to ration it. You can let a model think in the background for hours to solve a problem. Efficiency is the exact fuel that makes autonomous agency possible at scale.

The Agentic Era is now unfolding along two interesting paths. On one side, we have platforms like Claude Cowork. On the other side, we have the rise of the personal agent. OpenClaw is the prime example of this second path. OpenClaw is an open source framework that allows users to build and host their own autonomous agents. It acts like a digital employee that can run entirely on local hardware. It interacts with you through existing messaging applications and executes complex workflows autonomously. Because it does not rely on a corporate cloud, it keeps your data entirely private and creates a sovereign personal artificial intelligence.

The rise of OpenClaw also triggered a fascinating social experiment. Shortly after OpenClaw (it was then known as Moltbot) was created, Moltbook, a social network designed entirely for AI agents, was built. Humans are allowed only as observers who cannot post or comment. Today, over a million autonomous agents inhabit the platform. They generate posts, argue over technical concepts, share jokes, and upvote each other in a seemingly bizarre automated discourse. While it sounds like science fiction, it provides a striking glimpse into the future. When you give artificial intelligence the ability to act autonomously, it will eventually begin interacting with other machines. The internet is shifting from a human web to an agentic web.

This AI deluge leads us to the Jevons Paradox. In 1865, William Jevons observed that as steam engines became more efficient, coal consumption did not decline as expected. It instead rose rapidly because efficiency had made steam power viable for new uses. We are seeing the same thing today with AI. We will now use AI for tasks we previously considered too trivial to deploy AI. Humans will attempt to use AI in every aspect of our lives. Because the cost of thinking has collapsed, we can afford to let the AI agents think and do everything. The Jevons Paradox is driving the trillions of dollars in AI data center spending and the very real projects that envisage massive data centers under the sea or in space.

The New SaaS: Service-as-a-Software

We have spent more than 40 years teaching humans how to speak the language of computers. We learned to double-click icons, navigate drop-down menus, and type queries into search bars. We did this because computers were fast and capable, but did not think like us. They did not understand our world or our intent. We had to translate our intent into a series of clicks and taps that the machine could process. This necessity birthed the Graphical User Interface (GUI) and the App.

The Agentic Era marks the end of this compromise. The concept is simple but destructive to the status quo. In a world of intelligent agents, the App in its current form ceases to exist. Consider the daily reality of a B2B account executive managing a client renewal. Today, this process is manual across multiple screens and applications. The executive opens a CRM and the contract management system to check the contract terms, financials, and expiration date. They open a support portal to determine whether the client has any unresolved tickets. They open a dashboard to analyze product usage trends. Finally, they open their email to draft a message that attempts to synthesize these disparate data points. The human is the glue stitching these siloed pieces of software together.

In an Agentic AI future that is unfolding as we write this book, this workflow collapses into a single sentence of intent. "Prepare a renewal contract proposal for Acme Corp that factors in their usage in the last two quarters and their support tickets, offering a discount as per policy. Present it to me for approval before scheduling a review with their VP." The agent need not open a browser or look at a screen. It interacts directly with the APIs of the software systems. It negotiates the trade-offs and drafts the document. The user may never see the interface of all the underlying software.

This transition forces us to rethink the business model of the last two decades: Software-as-a-Service (SaaS). In the SaaS model, you buy a tool. You pay a subscription fee for a CRM tool like Salesforce or HubSpot. However, the tool sits idle until you hire a human to use it. The software is a lever, but the human provides the force. We are now moving to Service-as-a-Software, where you hire an AI agent. The software is no longer just a place to work. The software does the work. This is a challenge for software companies and will lead to a drastic change in the revenue model of the SaaS industry. As SaaS companies try to protect their turf and user interfaces, this will only hasten this change. Already, computer-use agents from OpenAI, Google, and Anthropic and the AI native browsers like OpenAI Atlas and Perplexity Comet are mimicking humans within these user interfaces, making it hard for systems to distinguish whether it is an AI agent or a human interacting with them. And they will only get better—really fast!

The Rise of the One-Person Unicorn

For the last decade, the math of building a startup was fairly rigid. If you wanted to build a company with a million dollars in revenue, you needed a certain number of salespeople, engineers, and support staff. A minimum viable headcount was required to build, sell, deploy, and maintain a minimum viable product. The average seed-stage startup had about six employees. To reach Unicorn status, a valuation of one billion dollars, companies typically needed hundreds of employees.

AI has broken this correlation. We are witnessing the rise of companies like Cursor, an AI-powered code editor, that achieved $100 million in annual recurring revenue in just 12 months with a small team of fewer than 20 people.

In the old world, a CEO sat atop a wide pyramid of management layers and individual contributors. In the new world, the CEO sits atop a much flatter organization, including a layer of autonomous agents. The founders of 2026 need to be architects of intelligence, skilled at defining intent and auditing the outcome. However, this new ease of creation brings a new danger: tech debt.

In December 2025, a debate erupted between Y Combinator's Garry Tan and Zoho's Sridhar Vembu on Twitter/X. Tan argued that "vibe coding," in which non-technical founders use AI to generate applications, would displace traditional SaaS. Why pay for software when you can speak it into existence? Vembu countered with the engineer's reality: maintenance. Vibe-coded apps work on day 1, but who fixes them on day 100? Who handles security, compliance, and edge cases? Both are right. Tan is right about the low end. Simple, internal tools may move from buy to build. Why pay for a generic tool when your operations agent can build a perfect one for five dollars in computing? But Vembu is right about the mission-critical layer. You will not use the vibe code for your core banking ledger. The risk is too high. The One-Person Unicorn will rise, but the successful ones will not just be prompt engineers. They will be system architects. They will be leaders who know when to trust the vibe and when they need to audit the code.

The Privacy Counter

We have spent the last two decades making a dangerous trade. We traded our privacy for convenience. We gave our location data to maps, our social connections to networks, and our search history to advertisers. In the Agentic Era, this trade goes up a notch. If an AI agent is going to be useful, it needs to know everything about you. It needs access to your medical records to schedule appointments, your bank accounts to pay bills, and your private emails to draft replies. If that agent lives in the cloud, owned by a mega-corporation, you have created the ultimate surveillance tool.

This is the privacy crisis of the next decade. We are witnessing the rise of Sovereign Personal AI. This is a model that does not live in a data center. It lives on your laptop and your phone. It runs locally on your own silicon. Before the DeepSeek moment, this was impossible. High-intelligence models were too large and unavailable for consumer hardware. You needed a server rack to run them.

DeepSeek changed the physics of deployment through Distillation. They demonstrated that you could take the intelligence of a large model and compress it into a smaller, efficient package. A model with 7 or 8 billion parameters can now reason at a level once requiring much larger systems, yet it runs on the memory chip of a standard smartphone. Many new laptops and phones now feature powerful, dedicated Neural Processing Units (NPUs) or AI engines for on-device AI workloads. This changes everything for privacy. Imagine a Health Agent. It analyzes your blood work, reads your genetic history, monitors your sleep data, and suggests lifestyle changes. You do not have to upload sensitive data to a cloud API. In the new world, the model comes to the data. The AI runs entirely on your device. Your medical history never leaves your pocket. Efficiency does not just save money. Efficiency protects freedom. You do not need to trust the cloud if the intelligence fits in your pocket.

Orchestrating the Federation

For the last three years, enterprise leaders have been hunting for the God Model. They wanted one single AI that could do it all. They wanted a model that could write code, draft legal briefs, translate French, and design marketing images. That search seems to be over. The future is not a monolith. It is a federation. In the Agentic Era, the most successful companies will not be those with the largest models. They will have the best team of specialized models and the best agents who know how to use them.

Remember the elephant on roller skates from Chapter 6? The challenge was teaching a massive, clumsy organization to move with speed. The Federated Architecture is how we keep the elephant upright. We do not try to replace the elephant's brain with one giant God Model. That is too risky. Instead, we give it specialized reflexes. The enterprise of 2030 will function as an octopus (surprised? The octopus has nine (yes, nine!) brains—one brain in its head has overall control, and the eight "ganglia" at the base of each arm act semi-independently for routine actions like tasting, touching, and moving) that allows its agents autonomous control, but orchestrates the outcomes for the enterprise in cooperation with human leadership.

The Ghosts of Hypes Past

We must distinguish between utility and mania. Artificial intelligence differs from many past technology bubbles. It is already delivering tremendous value. It is rewriting code, optimizing logistics, and accelerating research in ways that are tangible and measurable. However, a gap often exists between the validity of a technology and the scale of the hype surrounding it. History is full of moments where the excitement for an inevitable future ran far ahead of reality. Investors and founders often learn expensive lessons when they mistake a long-term trend for an immediate revolution.

A useful reminder comes from the fascination with blockchain in the early 2020s. The technology was brilliant, yet the gap between what was technically possible and what was operationally viable and useful for mainstream business remained stubbornly wide. Virtual Reality offers another more recent lesson. The conviction behind the Metaverse was so intense that Facebook renamed itself Meta to claim the future. Billions of dollars were poured into infrastructure for a digital world that felt just around the corner. Yet, that timeline proved elusive. The company recently announced significant cutbacks as the grand vision failed to materialize. AI risks walking a similar path. The current global build-out of data centers and energy infrastructure assumes a vertical line of adoption.

The DeepSeek moment has solved the cost of intelligence. It has not yet solved the reliability of intelligence. The current tools are powerful, but they require responsible deployment. Overconfidence can produce fragile systems. Misplaced faith in autonomous agents can create blind spots. If an agent performs correctly 99 percent of the time, it can still pose a serious and unacceptable risk in many banking workflows. We will likely see an Agentic Bubble. We will observe enterprise disasters in which unchecked agents cause financial or reputational losses because they lack the governance to handle edge cases.

AI will shape the next decade. The transformation is real, but it will not be smooth. The technologies that endure will be those that survive the scrutiny of real workflows, economic gravity, and messy human behavior. It is wise to pursue bold possibilities. It is equally wise to remember that technological history rewards the patient, the disciplined, and the grounded. The winners in this new era will be the ones who combine imagination with restraint.

The Next Frontier

We have argued throughout this book that efficiency is the new strategy. DeepSeek proved that smart architecture can beat brute force. However, the Jevons Paradox tells us that efficiency drives up consumption. As the cost of intelligence drops to near zero, demand will explode. We will move from millions of human queries to millions of autonomous agents running 24 hours a day. They will negotiate supply chains, optimize power grids, and conduct research while we sleep.

This agentic swarm will eventually consume the efficiency dividend that the DeepSeek moment provided. When that happens, we will hit a harder wall. We will hit the physical limits of silicon and perhaps the thermal limits of our planet. To sustain the Agentic Era over the long term, we will need to move beyond the current computing paradigm. We are already seeing the early signals of three breakthroughs that will define the post-DeepSeek world.

The Energy Moonshot: Nuclear Fusion

Current data centers are straining national power grids. The Stargate projects of today rely on conventional energy sources, but the future requires a cleaner, denser source. We are seeing early bets on Nuclear Fusion, the process that powers the sun. If successful, this changes the calculus of AI. Fusion offers the promise of nearly unlimited, carbon-free energy. It would allow us to run the massive, continuous training runs required for Artificial General Intelligence (AGI) without expanding our carbon footprint. There's a running joke that Nuclear Fusion is always a decade away. But many now think that, with the assistance of large AI models that enable research-grade intelligence, scientists may be able to turn it around faster than previously thought possible.

The Compute Leap: Quantum Computing

Google, Microsoft, and others are rapidly advancing Quantum Computing. They are developing quantum processors like Google's Willow and Microsoft's Majorana, designed to solve specific optimization problems that would take a classical supercomputer thousands of years. Like Nuclear Fusion, we are living in times where the theoretical and experimental work of decades has the potential to turn into reality much faster, with the assistance of AI and autonomous agents that accelerate the innovation process manifold.

The Thermal Escape: Data Centers in Space

Heat is the ultimate enemy of computing. Today, we cool servers with water and massive air conditioners. The logical, if extreme, conclusion is to move the heat where it does not matter. Radiate the heat into space. Recent moonshot discussions by Google, xAI, and others are exploring the feasibility of orbital data centers. In space, solar energy is abundant. The rapidly declining cost of space transportation has brought this idea within the realm of possibility.

These technologies are not your strategy for 2026. They are the infrastructure of the 2030s. They represent the return of brute force, but in a new form. DeepSeek enabled us to initiate the AI revolution. These technologies will enable us to complete it. For the strategic leader, the lesson is simple. Optimize for silicon today, but keep an eye on the physics of tomorrow.

The Open Garden

In Chapter 1, we described the artificial intelligence landscape as a Walled Garden. It was a fortress of scarcity. It was built on a foundation of billion-dollar data centers and guarded by a handful of technology giants. The prevailing belief was that high intelligence was a luxury good. It was available only to those who could afford the steep admission fee. The DeepSeek moment did not just lower the price of admission. It kicked a hole in the wall.

We are now witnessing the collapse of that fortress. We have moved from a world where we rationed reasoning to a world where we can use it like electricity. Because the cost of thinking has dropped to near zero, we will no longer reserve it for grand problems such as curing cancer or modeling climate change. We will also, widely and perhaps also indiscriminately, use it for the mundane. We will use it to debug code, negotiate bills, organize schedules, and tutor children.

For the leaders reading this book, the implications are profound. The garden is now open. The tools of creation are affordable and accessible. But accessibility does not guarantee success. The fact that anyone can build a bank does not mean everyone should.

The era of the gold rush for infrastructure is ending. The era of the gold rush for applications has begun. The constraints of capital are gone. They are replaced by the constraints of creativity and discipline.

Intelligence is abundant. The gatekeepers have been bypassed. The cost is near zero. What will you build?

Index

A

Accountability, 68, 74, 77

Adoption, 55

Agentic AI future, 92

Agentic benchmarks, 24, 25

Agentic Era, 90, 92, 94, 95, 97

AGI, *see* Artificial General
Intelligence (AGI)

Agility, 86

AI, *see* Artificial intelligence (AI)

AI hubs
China, 83
Europe, 85
the Gulf, 84
India, 83
Japan, 84
South Korea, 84

AI-powered code editor, 93

AI regulation, 70–73
China, 69
ethics, 73
EU, 68
high-risk AI, 68
limited-risk AI, 69
minimal-risk AI, 69
unacceptable-risk AI, 68
United Kingdom, 69
United States, 69

Anthropic, 25, 44, 89, 93

Anticipation premium (AP), 46, 47

Anti-fragility, 54, 57

AP, *see* Anticipation premium (AP)

"Arena" (Human-Preference
Leaderboards)
analogy, 27
flaw, 27
test drive, 27
value, 27

Artificial demand, 83

Artificial General Intelligence
(AGI), 3, 81, 98

Artificial intelligence (AI), 1, 31, 96
access zones
conditional middle, 82
restricted zone, 82
trusted core, 82
agents, 87, 90, 91, 94
anticipation premium
(AP), 46, 47
arms race, 4
business processes, 47
Custodians, 42
data, 5
democratization, 78
enterprise risk, 86, 87
enterprises, 50

G

H

I

J

K

L

M

N

O

U

V, W, X

Y

Z

GPSR Compliance
The European Union's (EU) General Product Safety Regulation (GPSR) is a set
of rules that requires consumer products to be safe and our obligations to
ensure this.

If you have any concerns about our products, you can contact us on

ProductSafety@springernature.com

In case Publisher is established outside the EU, the EU authorized
representative is:

Springer Nature Customer Service Center GmbH
Europaplatz 3
69115 Heidelberg, Germany